Soil Pollution: Origin, Monitoring and Remediation

Soil Pollution: Origin, Monitoring and Remediation

Sebastian Brown

Larsen & Keller
www.larsen-keller.com

Soil Pollution: Origin, Monitoring and Remediation
Sebastian Brown
ISBN: 978-1-64172-671-9 (Hardback)

▤ Larsen & Keller

Published by Larsen and Keller Education,
5 Penn Plaza,
19th Floor,
New York, NY 10001, USA

Cataloging-in-Publication Data

Soil pollution : origin, monitoring and remediation / Sebastian Brown.
    p. cm.
Includes bibliographical references and index.
ISBN 978-1-64172-671-9
1. Soil pollution. 2. Sewage disposal in the ground. 3. Soil protection.
4. Soil remediation. I. Brown, Sebastian.
TD878 .S65 2022
628.55--dc23

For more information regarding Larsen and Keller Education and its products, please visit the publisher's website www.larsen-keller.com

Table of Contents

Preface

The degradation of soil by an alteration in the natural soil environment or through the presence of xenobiotic chemicals is known as soil pollution. It is usually caused by improper disposal of waste, agricultural chemicals and industrial activity. Some of the prominent chemicals which cause soil pollution are petroleum hyrdocarbons, naphthalene, solvents, pesticides and heavy metals such as lead. The process of removing pollutants and contaminants from the soil is known as environmental remediation. There are various strategies which are used in remediation. Some of these are aeration of soils, bioremediation, extracting groundwater and then filtering the contaminants, phytoremediation and surfactant leaching. The topics covered in this extensive book deal with the core aspects of soil pollution as well as its monitoring and remediation. It will serve as a reference to a broad spectrum of readers. Coherent flow of topics, student-friendly language and extensive use of examples make this book an invaluable source of knowledge.

A foreword of all Chapters of the book is provided below:

Chapter 1 - Soil pollution is defined as the contamination and degradation of soil due to the presence of pollutants and toxic chemicals in high concentrations. Pollutants such as lead, mercury, arsenic, copper, zinc, nickel, etc. pose a threat to the human health, wildlife and the environment. This chapter discusses the diverse aspects of soil pollution and its pollutants in detail; **Chapter 2** - Soil pollution is caused by various natural and man-made factors which include acid rain, overgrazing, urbanisation, landfill and illegal dumping, mining, use of agrochemicals and petrochemicals, etc. This chapter has been carefully written to provide an easy understanding of these different causes of soil pollution; **Chapter 3** - Soil pollution monitoring and assessment refers to the prevention, management, risk assessment and mitigation of the toxic agents of soil that adversely affect the land, air, water and living organisms. The topics elaborated in this chapter will help in gaining a better perspective about soil pollution monitoring and assessment; **Chapter 4** - There are different techniques of soil pollution treatment and remediation. Some of these are excavation and off-site disposal, on-site natural attenuation, bioslurping, bioventing, biosparging, phytoremediation, etc. This chapter closely examines these soil pollution treatment and control techniques to provide an extensive understanding of the subject; **Chapter 5** - Soil pollution adversely affects the wildlife, humans and the environment. It causes diseases like leukaemia, nervous system damage, etc. in humans, destruction of wildlife and its habitat, impact on soil fertility and agriculture, etc. All these impacts of soil pollution have been carefully analysed in this chapter.

I would like to thank the entire editorial team who made sincere efforts for this book and my family who supported me in my efforts of working on this book. I take this opportunity to thank all those who have been a guiding force throughout my life.

Sebastian Brown

An Introduction to Soil Pollution

Soil pollution is defined as the contamination and degradation of soil due to the presence of pollutants and toxic chemicals in high concentrations. Pollutants such as lead, mercury, arsenic, copper, zinc, nickel, etc. pose a threat to the human health, wildlife and the environment. This chapter discusses the diverse aspects of soil pollution and its pollutants in detail.

Soil Pollution

Soil pollution refers to anything that causes contamination of soil and degrades the soil quality. It occurs when the pollutants causing the pollution reduce the quality of the soil and convert the soil inhabitable for microorganisms and macro organisms living in the soil.

Soil contamination or soil pollution can occur either because of human activities or because of natural processes. However, mostly it is due to human activities. The soil contamination can occur due to the presence of chemicals such as pesticides, herbicides, ammonia, petroleum hydrocarbons, lead, nitrate, mercury, naphthalene, etc in an excess amount.

The primary cause of soil pollution is a lack of awareness in general people. Thus, due to many different human activities such as overuse of pesticides the soil will lose its fertility. Moreover, the presence of excess chemicals will increase the alkalinity or acidity of soil thus degrading the soil quality. This will in turn cause soil erosion. This soil erosion refers to soil pollution.

The problem of soil pollution arises due to mixing of toxic and polluted materials in the soil. Illegal dumping is the biggest reason for soil pollution, which adversely affects the quality of soil and the health of people living on it. Soil pollution also spreads through polluted water absorbed by the soil. Chemical compost used in agricultural work, litter and dirt also badly pollute the soil. The soil is also polluted by the mineral oil spread on the land accidently. Pollutants present in the air also contribute to polluting the soil. Through the rain water, pollutants present in the air descend on the ground which ultimately results into polluting the soil.

Soil is an important natural resource on Earth that is essential to run the life of humankind and animals by producing vegetation, grains and other natural substances required for food and living. Fertile soil on Earth is essential for the production of crops which is essential for the food of all living beings. Fertility of the land is severely affected due to the inclusion of toxic elements in the soil due to chemical fertilizers, pesticides, and industrial effluents.

Let us take into the accounts some statistics. Between the years 1999 to the farmers worldwide used 18.07 million tonnes of chemical fertilizers and the use of chemical fertilizers is still going on uninterruptedly. These toxic chemicals pollute the soil and ultimately enter the food chain and infect us with dangerous diseases. Even the newborn babies and infants take birth with many types of physical inefficiency due to this phenomenon.

About 1000 square miles of land in Tacoma, Washington had been polluted in minutes due to airborne pollutants falling on the ground; hence the incident is cited as a grave instance of soil pollution.

Harmful chemicals present in the air pour down as acids in the form of rain and contribute to raising the soil pollution to dangerous levels. The direct effect of polluted soil is on the health of men and animals. The crop produced in the soil polluted by harmful chemical substances causes cancer and other incurable diseases by reaching the body of humans and other living organisms.

Due to the large-scale industrialization, industrial effluents are continuously discharged in wastewater. As a result, heavy metals are mixed in soil, turning it toxic.

World scientists have warned from time to time that if there is no attention given on time to soil pollution, it can lead to disastrous consequences. Due to polluted soil, there is an adverse effect on the yield of crops too. Along with the other countries, thousands of hectares of agricultural land in many parts of India have lost their fertility through the constant use of chemical fertilizers.

Types of Soil Pollution

There are many types of natural- and human-borne soil pollution:

Land Pollution from Domestic and Industrial Solid Waste

Electronic goods, broken furniture, junk papers, polythene bags, plastic cans, bottles, wastewater, toxic waste from the hospital etc. are examples of solid waste which pollute the soil. Most of this litter is non biodegradable. These wastes affect the soil structure by being blocked in it for long periods. Because these solid wastes do not decay easily, they lie on landfill sites for thousands of years and keep polluting the soil and the environment continuously. In addition to the soil, humans and animals living around these landfill sites are greatly harmed.

Household waste, industrial waste etc. contain residues of harmful toxic inorganic and organic chemicals. In these residues, radiation elements such as strontium, cadmium, uranium, ladders are found, which affect the vitality and fertility of the land. Fly ash is a major source of pollution surrounding the industrial area.

There are chemicals or other types of waste in industries, which are dumped at some place. So much so that soil becomes polluted and trees and plants do not even grow in such a part.

Soil Pollution by Chemical Substances

The use of chemical pesticides and fertilizers has increased for cultivating more crops and these pollutants are making the soil poisonous and in many places the soil has become dead due to excessive use of it.

Producers of fertilizers, insecticides, pesticides, pharmaceuticals produce a lot of solid and liquid waste. Due to leaks from pipes and gutters, pollutants also go into the soil and spread pollution.

In the chemical and nuclear power plants, a large amount of waste is released continuously and due to the absence of proper arrangements for their storage and disposal, these substances pollute the soil.

In commercial agriculture, insecticides are being used indiscriminately and inorganic chemical fertilizers are also being used day by day. The chemical fertilizers are polluting the environment and groundwater resources of phosphate, nitrogen and other organic chemical land. The most dangerous pollutants are bioactive chemicals, due to which the micro-organisms of climates and other soil are being destroyed resulting in decreased quality of soil. Toxic chemicals enter the diet chain, so that they reach the top consumer. Bioactive chemicals are also called Creeping Deaths. In the last 30 years, the use of organic chemicals has increased by more than 11 times. India alone is using 100,000 tonnes of bio-chemicals per annum.

Continuous Deforestation

Trees absorb carbon dioxide from the air; provide oxygen for humans and other organisms. Apart from these, tree plantations are also helpful in prevention of soil pollution and erosion. Tree plantation rejuvenates the lost potency of soil. But unfortunately, we are continuously cutting trees on the millions of acres of land for the wood required for construction and the land required for the cultivation, besides mining work.

Factors Responsible for Soil Pollution at a Glance

There are many factors responsible for soil pollution such as chemical substances and insecticides, oil spills, landfill dumps and industrial waste etc. Pesticides used in agricultural work are mixed in the air and acid is found in the form of rain. Due to the rapid erosion of forests, the soil has been badly affected by pollution in the world.

Here are prime factors responsible for soil pollution at a glance:

- Continuous drilling in mineral oil and oil wells.

- Mining activities to achieve the essential minerals need to run heavy industries. Wreckage from mining is put in a nearby place. Debris from the excavation of minerals such as stone, iron, ore, mica, copper, etc., eliminates the fertile power of the soil. Together with the water at the time of the rain, the debris goes far away and pollutes the soil.

- Accidents arising during mining activities such as accidents due to accident from oil wells, expansion of oil on land, or during the mining activity for obtaining uranium etc.

- Leakage from pipes being used to transmit oil to the refining plants by the tanks being made for underground oil storage.

- Acid rain carries dangerous levels of pollutants in the air.

- Use of chemical fertilizers to get more crops during agricultural work.

- Industrial accidents due to which hazardous chemicals are mixed in the soil.

- Roads and places where debris is deployed.

- Dehydration of contaminated water in soil.

- Soil disposal of waste, oil and fuels.

- Disposal of atomic waste.

- Construction of landfill and illegal dumping spots.

- Ashes born after burning coal.

- Large amount of electronic waste production.

Main Causes of Soil Pollution

Industrial Activity

Industrial activity has been the biggest contributor to the problem in the last century, especially since the amount of mining and manufacturing has increased. Most industries are dependent on extracting minerals from the Earth. Whether it is iron ore or coal, the by-products are contaminated and they are not disposed of in a manner that can be considered safe. As a result, the industrial waste lingers in the soil surface for a long time and makes it unsuitable for use.

Agricultural Activities

Chemical utilization has gone up tremendously since technology provided us with modern pesticides and fertilizers. They are full of chemicals that are not produced in nature and cannot be broken down by it. As a result, they seep into the ground after they mix with water and slowly reduce the fertility of the soil.

Other chemicals damage the composition of the soil and make it easier to erode by water and air. Plants absorb many of these pesticides and when they decompose, they cause soil pollution since they become a part of the land.

Waste Disposal

Finally, a growing cause for concern is how we dispose of our waste. While industrial waste is sure to cause contamination, there is another way in which we are adding to the pollution. Every human produces a certain amount of personal waste products by way or urine and feces.

While much of it moves into the sewer the system, there is also a large amount that is dumped directly into landfills in the form of diapers. Even the sewer system ends at the landfill, where the biological waste pollutes the soil and water. This is because our bodies are full of toxins and chemicals which are now seeping into the land and causing pollution of soil.

Accidental Oil Spills

Oil leaks can happen during storage and transport of chemicals. This can be seen at most of the fuel stations. The chemicals present in the fuel deteriorates the quality of soil and make them unsuitable for cultivation. These chemicals can enter into the groundwater through the soil and make the water undrinkable.

Acid Rain

Acid rain is caused when pollutants present in the air mix up with the rain and fall back on the ground. The polluted water could dissolve away some of the important nutrients found in soil and change the structure of the soil.

Effects of Soil Pollution

Effect on Health of Humans

Considering how soil is the reason we are able to sustain ourselves, the contamination of it has major consequences on our health. Crops and plants are grown on polluted soil absorb much of the pollution and then pass these on to us. This could explain the sudden surge in small and terminal illnesses.

Long term exposure to such soil can affect the genetic make-up of the body, causing congenital illnesses and chronic health problems that cannot be cured easily. In fact, it can sicken the livestock to a considerable extent and cause food poisoning over a long period of time. The soil pollution can even lead to widespread famines if the plants are unable to grow in it.

Effect on Growth of Plants

The ecological balance of any system gets affected due to the widespread contamination of the soil. Most plants are unable to adapt when the chemistry of the soil changes so radically in a short period of time. Fungi and bacteria found in the soil that bind it together begin to decline, which creates an additional problem of soil erosion.

The fertility slowly diminishes, making land unsuitable for agriculture and any local vegetation to survive. The soil pollution causes large tracts of land to become hazardous to health. Unlike deserts, which are suitable for its native vegetation, such land cannot support most forms of life.

Decreased Soil Fertility

The toxic chemicals present in the soil can decrease soil fertility and therefore decrease in the soil yield. The contaminated soil is then used to produce fruits and vegetables which lacks quality nutrients and may contain some poisonous substance to cause serious health problems in people consuming them.

Toxic Dust

The emission of toxic and foul gases from landfills pollutes the environment and causes serious effects on the health of some people. The unpleasant smell causes inconvenience to other people.

Changes in Soil Structure

The death of many soil organisms (e.g. earthworms) in the soil can lead to alteration in soil structure. Apart from that, it could also force other predators to move to other places in search of food.

Soil Pollutants

Soil pollutant consider as any chemical of natural or anthropogenic origin which accumulates in the soil medium and changes the natural soil equilibrium, as a result of human activity. Soil pollutants can be included into two main groups, the organic pollutants (OPs) and the inorganic pollutants (IPs).

Some of the most hazardous soil pollutants are xenobiotics – substances that are not naturally found in nature and are synthesized by human beings. Several xenobiotics are known to be carcinogens.

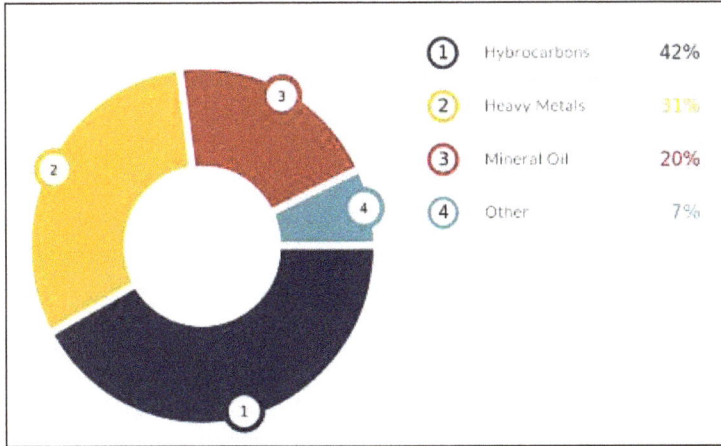

(1)	Hybrocarbons	42%
(2)	Heavy Metals	31%
(3)	Mineral Oil	20%
(4)	Other	7%

Heavy Metals

The presence of heavy metals (such as lead and mercury, in abnormally high concentrations) in soils can cause it to become highly toxic to human beings. Some metals that can be classified as soil pollutants are tabulated below.

Toxic Metals that Cause Soil Pollution		
Arsenic	Mercury	Lead
Antimony	Zinc	Nickel
Cadmium	Selenium	Beryllium
Thallium	Chromium	Copper

These metals can originate from several sources such as mining activities, agricultural activities, electronic waste (e-waste), and medical waste.

Polycyclic Aromatic Hydrocarbons

Polycyclic aromatic hydrocarbons (often abbreviated to PAHs) are organic compounds that:

- Contain only carbon and hydrogen atoms.
- Contain more than one aromatic ring in their chemical structures.

Common examples of PAHs include naphthalene, anthracene, and phenalene. Exposure to polycyclic aromatic hydrocarbons has been linked to several forms of cancer. These organic compounds can also cause cardiovascular diseases in humans.

Soil pollution due to PAHs can be sourced to coke (coal) processing, vehicle emissions, cigarette smoke, and the extraction of shale oil.

Industrial Waste

The discharge of industrial waste into soils can result in soil pollution. Some common soil pollutants that can be sourced to industrial waste are listed below:

- Chlorinated industrial solvents.
- Dioxins produced from the manufacture of pesticides and the incineration of waste.
- Plasticizers/dispersants.
- Polychlorinated biphenyls (PCBs).

The petroleum industry creates many petroleum hydrocarbon waste products. Some of these wastes, such as benzene and methylbenzene, are known to be carcinogenic in nature.

Pesticides

Pesticides are substances (or mixtures of substances) that are used to kill or inhibit the growth of pests. Common types of pesticides used in agriculture include:

- Herbicides – used to kill/control weeds and other unwanted plants.
- Insecticides – used to kill insects.
- Fungicides – used to kill parasitic fungi or inhibit their growth.

However, the unintentional diffusion of pesticides into the environment (commonly known as 'pesticide drift') poses a variety of environmental concerns such as water pollution and soil pollution. Some important soil contaminants found in pesticides are listed below.

Herbicides

- Triazines,
- Carbamates,
- Amides,
- Phenoxyalkyl acids,
- Aliphatic acids.

Insecticides

- Organophosphates,
- Chlorinated hydrocarbons,

- Arsenic-containing compounds,

- Pyrethrum.

Fungicides

- Mercury-containing compounds,

- Thiocarbamates,

- Copper sulfate.

These chemicals pose several health risks to humans. Examples of health hazards related to pesticides include diseases of the central nervous system, immune system diseases, cancer, and birth defects.

Soil Polluted by Heavy Metals

Soils polluted with heavy metals have become common across the globe due to increase in geologic and anthropogenic activities. Plants growing on these soils show a reduction in growth, performance, and yield. Bioremediation is an effective method of treating heavy metal polluted soils. It is a widely accepted method that is mostly carried out in situ; hence it is suitable for the establishment/reestablishment of crops on treated soils. Microorganisms and plants employ different mechanisms for the bioremediation of polluted soils. Using plants for the treatment of polluted soils is a more common approach in the bioremediation of heavy metal polluted soils. Combining both microorganisms and plants is an approach to bioremediation that ensures a more efficient clean-up of heavy metal polluted soils. However, success of this approach largely depends on the species of organisms involved in the process.

Although heavy metals are naturally present in the soil, geologic and anthropogenic activities increase the concentration of these elements to amounts that are harmful to both plants and animals. Some of these activities include mining and smelting of metals, burning of fossil fuels, use of fertilizers and pesticides in agriculture, production of batteries and other metal products in industries, sewage sludge, and municipal waste disposal.

Growth reduction as a result of changes in physiological and biochemical processes in plants growing on heavy metal polluted soils has been recorded. Continued decline in plant growth reduces yield which eventually leads to food insecurity. Therefore, the remediation of heavy metal polluted soils cannot be overemphasized.

Various methods of remediating metal polluted soils exist; they range from physical and chemical methods to biological methods. Most physical and chemical methods (such as encapsulation, solidification, stabilization, electrokinetics, vitrification, vapour extraction, and soil washing and flushing) are expensive and do not make the soil suitable for plant growth. Biological approach (bioremediation) on the other hand encourages the establishment/reestablishment of plants on polluted soils. It is an environmentally friendly approach because it is achieved via natural processes. Bioremediation is also an economical remediation technique compared with other remediation techniques.

Heavy Metal Polluted Soils

Heavy metals are elements that exhibit metallic properties such as ductility, malleability, conductivity, cation stability, and ligand specificity. They are characterized by relatively high density and high relative atomic weight with an atomic number greater than 20. Some heavy metals such as Co, Cu, Fe, Mn, Mo, Ni, V, and Zn are required in minute quantities by organisms. However, excessive amounts of these elements can become harmful to organisms. Other heavy metals such as Pb, Cd, Hg, and As (a metalloid but generally referred to as a heavy metal) do not have any beneficial effect on organisms and are thus regarded as the "main threats" since they are very harmful to both plants and animals.

Metals exist either as separate entities or in combination with other soil components. These components may include exchangeable ions sorbed on the surfaces of inorganic solids, nonexchangeable ions and insoluble inorganic metal compounds such as carbonates and phosphates, soluble metal compound or free metal ions in the soil solution, metal complex of organic materials, and metals attached to silicate minerals. Metals bound to silicate minerals represent the background soil metal concentration and they do not cause contamination/pollution problems compared with metals that exist as separate entities or those present in high concentration in the other 4 components.

Soil properties affect metal availability in diverse ways. Harter reported that soil pH is the major factor affecting metal availability in soil. Availability of Cd and Zn to the roots of Thlaspi caerulescens decreased with increases in soil pH. Organic matter and hydrous ferric oxide have been shown to decrease heavy metal availability through immobilization of these metals. Significant positive correlations have also been recorded between heavy metals and some soil physical properties such as moisture content and water holding capacity.

Other factors that affect the metal availability in soil include the density and type of charge in soil colloids, the degree of complexation with ligands, and the soil's relative surface area. The large interface and specific surface areas provided by soil colloids help in controlling the concentration of heavy metals in natural soils. In addition, soluble concentrations of metals in polluted soils may be reduced by soil particles with high specific surface area, though this may be metal specific. For instance, Mcbride and Martínez reported that addition of amendment consisting of hydroxides with high reactive surface area decreased the solubility of As, Cd, Cu, Mo, and Pb while the solubility of Ni and Zn was not changed. Soil aeration, microbial activity, and mineral composition have also been shown to influence heavy metal availability in soils.

Conversely, heavy metals may modify soil properties especially soil biological properties. Monitoring changes in soil microbiological and biochemical properties after contamination can be used to evaluate the intensity of soil pollution because these methods are more sensitive and results can be obtained at a faster rate compared with monitoring soil physical and chemical properties. Heavy metals affect the number, diversity, and activities of soil microorganisms. The toxicity of these metals on microorganisms depends on a number of factors such as soil temperature, pH, clay minerals, organic matter, inorganic anions and cations, and chemical forms of the metal.

There are discrepancies in studies comparing the effect of heavy metals on soil biological properties. While some researchers have recorded negative effect of heavy metals on soil biological properties, others have reported no relationship between high heavy metal concentrations and some soil (micro) biological properties. Some of the inconsistencies may arise because some of these studies were conducted under laboratory conditions using artificially contaminated soils while others were carried out using soils from areas that are actually polluted in the field. Regardless of the origin of the soils used in these experiments, the fact that the effect of heavy metals on soil biological properties needs to be studied in more detail in order to fully understand the effect of these metals on the soil ecosystem remains. Further, it is advisable to use a wide range of methods (such as microbial biomass, C and N mineralization, respiration, and enzymatic activities) when studying effect of metals on soil biological properties rather than focusing on a single method since results obtained from use of different methods would be more comprehensive and conclusive.

The presence of one heavy metal may affect the availability of another in the soil and hence plant. In other words, antagonistic and synergistic behaviours exist among heavy metals. Salgare and Acharekar reported that the inhibitory effect of Mn on the total amount of mineralized C was antagonized by the presence of Cd. Similarly, Cu and Zn as well as Ni and Cd have been reported to compete for the same membrane carriers in plants. In contrast, Cu was reported to increase the toxicity of Zn in spring barley. This implies that the interrelationship between heavy metals is quite complex; thus more research is needed in this area.

Effect of Heavy Metal Polluted Soil on Plant Growth

The heavy metals that are available for plant uptake are those that are present as soluble components in the soil solution or those that are easily solubilized by root exudates. Although plants require certain heavy metals for their growth and upkeep, excessive amounts of these metals can become toxic to plants. The ability of plants to accumulate essential metals equally enables them to acquire other nonessential metals. As metals cannot be broken down, when concentrations within the plant exceed optimal levels, they adversely affect the plant both directly and indirectly.

Some of the direct toxic effects caused by high metal concentration include inhibition of cytoplasmic enzymes and damage to cell structures due to oxidative stress. An example of indirect toxic effect is the replacement of essential nutrients at cation exchange sites of plants. Further, the negative influence heavy metals have on the growth and activities of soil microorganisms may also indirectly affect the growth of plants. For instance, a reduction in the number of beneficial soil microorganisms due to high metal concentration may lead to decrease in organic matter decomposition leading to a decline in soil nutrients. Enzyme activities useful for plant metabolism may also be hampered due to heavy metal interference with activities of soil microorganisms. These toxic effects (both direct and indirect) lead to a decline in plant growth which sometimes results in the death of the plant.

The effect of heavy metal toxicity on the growth of plants varies according to the particular heavy metal involved in the process. Here is a detail of the toxic effects of specific metals on growth, biochemistry, and physiology of various plants. For metals such as Pb, Cd, Hg, and As which do not play any beneficial role in plant growth, adverse effects have been recorded at very low concentrations of these metals in the growth medium. Kibra recorded significant reduction in height of rice plants growing on a soil contaminated with 1 mg Hg/kg. Reduced tiller and panicle formation also occurred at this concentration of Hg in the soil. For Cd, reduction in shoot and root growth in wheat plants occurred when Cd in the soil solution was as low as 5 mg/L. Most of the reduction in growth parameters of plants growing on polluted soils can be attributed to reduced photosynthetic activities, plant mineral nutrition, and reduced activity of some enzymes.

Effect of Heavy Metal Toxicity on Plants

Heavy Metal: As

Rice (Oryza sativa) - Reduction in seed germination; decrease in seedling height; reduced leaf area and dry matter production.

Tomato (Lycopersicon esculentum) - Reduced fruit yield; decrease in leaf fresh weight.

Canola (Brassica napus) - Stunted growth; chlorosis; wilting.

Heavy Metal: Cd

Wheat (Triticum sp.) - Reduction in seed germination; decrease in plant nutrient content; reduced shoot and root length.

Garlic (Allium sativum) - Reduced shoot growth; Cd accumulation.

Maize (Zea mays) - Reduced shoot growth; inhibition of root growth.

Heavy Metal: Co

Tomato (Lycopersicon esculentum) - Reduction in plant nutrient content.

Mung bean (Vigna radiata) - Reduction in antioxidant enzyme activities; decrease in plant sugar, starch, amino acids, and protein content.

Radish (Raphanus sativus) - Reduction in shoot length, root length, and total leaf area; decrease in chlorophyll content; reduction in plant nutrient content and antioxidant enzyme activity; decrease in plant sugar, amino acid, and protein content.

Heavy Metal: Cr

Wheat (Triticum sp.) - Reduced shoot and root growth.

Tomato (Lycopersicon esculentum) - Decrease in plant nutrient acquisition.

Onion (Allium cepa) - Inhibition of germination process; reduction of plant biomass.

Heavy Metal: Cu

Bean (Phaseolus vulgaris) - Accumulation of Cu in plant roots; root malformation and reduction.

Black bindweed (Polygonum convolvulus) - Plant mortality; reduced biomass and seed production.

Rhodes grass (Chloris gayana) - Root growth reduction.

Heavy Metal: Hg

Rice (Oryza sativa) - Decrease in plant height; reduced tiller and panicle formation; yield reduction; bioaccumulation in shoot and root of seedlings.

Tomato (Lycopersicon esculentum) - Reduction in germination percentage; reduced plant height; reduction in flowering and fruit weight; chlorosis.

Heavy Metal: Mn

Broad bean (Vicia faba) - Mn accumulation shoot and root; reduction in shoot and root length; chlorosis.

Spearmint (Mentha spicata) - Decrease in chlorophyll a and carotenoid content; accumulation of Mn in plant roots.

Pea (Pisum sativum) - Reduction in chlorophylls a and b content; reduction in relative growth rate; reduced photosynthetic O_2 evolution activity and photosystem II activity.

Tomato (Lycopersicon esculentum) - Slower plant growth; decrease in chlorophyll concentration.

Heavy Metal: Ni

Pigeon pea (Cajanus cajan) - Decrease in chlorophyll content and stomatal conductance; decreased enzyme activity which affected Calvin cycle and CO_2 fixation.

Rye grass (Lolium perenne) - Reduction in plant nutrient acquisition; decrease in shoot yield; chlorosis.

Wheat (Triticum sp.) - Reduction in plant nutrient acquisition.

Rice (Oryza sativa) - Inhibition of root growth.

Heavy Metal: Pb

Maize (Zea mays) - Reduction in germination percentage; suppressed growth; reduced plant biomass; decrease in plant protein content.

Portia tree (Thespesia populnea) - Reduction in number of leaves and leaf area; reduced plant height; decrease in plant biomass.

Oat (Avena sativa) - Inhibition of enzyme activity which affected CO_2 fixation.

Heavy Metal: Zn

Cluster bean (Cyamopsis tetragonoloba) - Reduction in germination percentage; reduced plant height and biomass; decrease in chlorophyll, carotenoid, sugar, starch, and amino acid content.

Pea (Pisum sativum) - Reduction in chlorophyll content; alteration in structure of chloroplast; reduction in photosystem II activity; reduced plant growth.

Rye grass (Lolium perenne) - Accumulation of Zn in plant leaves; growth reduction; decrease in plant nutrient content; reduced efficiency of photosynthetic energy conversion.

For other metals which are beneficial to plants, "small" concentrations of these metals in the soil could actually improve plant growth and development. However, at higher concentrations of these metals, reductions in plant growth have been recorded. For instance, Jayakumar et al. reported that, at 50 mg Co/kg, there was an increase in nutrient content of tomato plants compared with the control. Conversely, at 100 mg Co/kg to 250 mg Co/kg, reductions in plant nutrient content were recorded. Similarly, increase in plant growth, nutrient content, biochemical content, and antioxidant enzyme activities (catalase) was observed in radish and mung bean at 50 mg Co/kg soil concentration while reductions were recorded at 100 mg Co/kg to 250 mg Co/kg soil concentration. Improvements in growth and physiology of cluster beans have also been reported at Zn concentration of 25 mg/L of the soil solution. On the other hand, growth reduction and adverse effect on the plant's physiology started when the soil solution contained 50 mg Zn/L.

It is worth mentioning that, in most real life situations (such as disposal of sewage sludge and metal mining wastes) where soil may be polluted with more than one heavy metal, both antagonistic and synergistic relationships between heavy metals may affect plant metal toxicity. Nicholls and Mal reported that the combination of Pb and Cu at both high concentration (1000 mg/kg each) and low concentration (500 mg/kg) resulted in a rapid and complete death of the leaves and stem of Lythrum salicaria.

Sources of Heavy Metal in Polluted Soil

Heavy metals occur naturally in the soil environment from the pedogenetic processes of weathering of parent materials at levels that are regarded as trace ($<1000\,mg\,kg^{-1}$) and rarely toxic. Due to the disturbance and acceleration of nature's slowly occurring geochemical cycle of metals by man, most soils of rural and urban environments may accumulate one or more of the heavy metals above defined background values high enough to cause risks to human health, plants, animals, ecosystems, or other media. The heavy metals essentially become contaminants in the soil environments because (i) their rates of generation via man-made cycles are more rapid relative to natural ones, (ii) they become transferred from mines to random environmental locations where higher potentials of direct exposure occur, (iii) the concentrations of the metals in discarded products are relatively high compared to those in the receiving environment, and (iv) the chemical form (species) in which a metal is found in the receiving environmental system may render it more bioavailable. A simple mass balance of the heavy metals in the soil can be expressed as follows:

$$M_{total} = \left(M_p + M_a + M_f + M_{ag} + M_{ow} + M_{ip} \right) - \left(M_{cr} + M_l \right),$$

where "M" is the heavy metal, "p" is the parent material, "a" is the atmospheric deposition, "f" is the fertilizer sources, "ag" are the agrochemical sources, "ow" are the organic waste sources, "ip" are other inorganic pollutants, "cr" is crop removal, and "l" is the losses by leaching, volatilization, and so forth. It is projected that the anthropogenic emission into the atmosphere, for several heavy metals, is one-to-three orders of magnitude higher than natural fluxes. Heavy metals in the soil from anthropogenic sources tend to be more mobile, hence bioavailable than pedogenic, or lithogenic ones. Metal-bearing solids at contaminated sites can originate from a wide variety of anthropogenic sources in the form of metal mine tailings, disposal of high metal wastes in improperly protected landfills, leaded gasoline and lead-based paints, land application of fertilizer, animal manures, biosolids (sewage sludge), compost, pesticides, coal combustion residues, petrochemicals, and atmospheric deposition are discussed hereunder.

Fertilizers

Historically, agriculture was the first major human influence on the soil. To grow and complete the lifecycle, plants must acquire not only macronutrients (N, P, K, S, Ca, and Mg), but also essential micronutrients. Some soils are deficient in the heavy metals (such as Co, Cu, Fe, Mn, Mo, Ni, and Zn) that are essential for healthy plant growth, and crops may be supplied with these as an addition to the soil or as a foliar spray. Cereal crops grown on Cu-deficient soils are occasionally treated with Cu as an addition to the soil, and Mn may similarly be supplied to cereal and root crops. Large quantities of fertilizers are regularly added to soils in intensive farming systems to provide adequate N, P, and K for crop growth. The compounds used to supply these elements contain trace amounts of heavy metals (e.g., Cd and Pb) as impurities, which, after continued fertilizer, application may significantly increase their content in the soil. Metals, such as Cd and Pb, have no known physiological activity. Application of certain phosphatic fertilizers inadvertently adds Cd and other potentially toxic elements to the soil, including F, Hg, and Pb.

Pesticides

Several common pesticides used fairly extensively in agriculture and horticulture in the past contained substantial concentrations of metals. For instance in the recent past, about 10% of the chemicals have approved for use as insecticides and fungicides in UK were based on compounds which contain Cu, Hg, Mn, Pb, or Zn. Examples of such pesticides are copper-containing fungicidal sprays such as Bordeaux mixture (copper sulphate) and copper oxychloride. Lead arsenate was used in fruit orchards for many years to control some parasitic insects. Arsenic-containing compounds were also used extensively to control cattle ticks and to control pests in banana in New Zealand and Australia, timbers have been preserved with formulations of Cu, Cr, and As (CCA), and there are now many derelict sites where soil concentrations of these elements greatly exceed background concentrations. Such contamination has the potential to cause problems, particularly if sites are redeveloped for other agricultural or nonagricultural purposes. Compared with fertilizers, the use of such materials has been more localized, being restricted to particular sites or crops.

Biosolids and Manures

The application of numerous biosolids (e.g., livestock manures, composts, and municipal sewage sludge) to land inadvertently leads to the accumulation of heavy metals such as As, Cd, Cr, Cu, Pb, Hg, Ni, Sc, Mo, Zn, Tl, Sb, and so forth, in the soil. Certain animal wastes such as poultry, cattle, and pig manures produced in agriculture are commonly applied to crops and pastures either as solids or slurries. Although most manures are seen as valuable fertilizers, in the pig and poultry industry, the Cu and Zn added to diets as growth promoters and As contained in poultry health products may also have the potential to cause metal contamination of the soil. The manures produced from animals on such diets contain high concentrations of As, Cu, and Zn and, if repeatedly applied to restricted areas of land, can cause considerable buildup of these metals in the soil in the long run.

Biosolids (sewage sludge) are primarily organic solid products, produced by wastewater treatment processes that can be beneficially recycled. Land application of biosolids materials is a common practice in many countries that allow the reuse of biosolids produced by urban populations. The term sewage sludge is used in many references because of its wide recognition and its regulatory definition. However, the term biosolids is becoming more common as a replacement for sewage sludge because it is thought to reflect more accurately the beneficial characteristics inherent to sewage sludge. It is estimated that in the United States, more than half of approximately 5.6 million dry tonnes of sewage sludge used or disposed of annually is land applied, and agricultural utilization of biosolids occurs in every region of the country. In the European community, over 30% of the sewage sludge is used as fertilizer in agriculture. In Australia over 175 000 tonnes of dry biosolids are produced each year by the major metropolitan authorities, and currently most biosolids applied to agricultural land are used in arable cropping situations where they can be incorporated into the soil.

There is also considerable interest in the potential for composting biosolids with other organic materials such as sawdust, straw, or garden waste. If this trend continues, there will be implications for metal contamination of soils. The potential of biosolids for contaminating soils with heavy metals has caused great concern about their application in agricultural practices. Heavy metals most commonly found in biosolids are Pb, Ni, Cd, Cr, Cu, and Zn, and the metal concentrations are governed by the nature and the intensity of the industrial activity, as well as the type of process employed during the biosolids treatment. Under certain conditions, metals added to soils in applications of biosolids can be leached downwards through the soil profile and can have the potential to contaminate groundwater. Recent studies on some New Zealand soils treated with biosolids have shown increased concentrations of Cd, Ni, and Zn in drainage leachates.

Wastewater

The application of municipal and industrial wastewater and related effluents to land dates back 400 years and now is a common practice in many parts of the world. Worldwide, it is estimated that 20 million hectares of arable land are irrigated with waste water. In several Asian and African cities, studies suggest that agriculture based on wastewater irrigation accounts for 50 percent of the vegetable supply to urban areas. Farmers generally are not bothered about environmental benefits or hazards and are primarily interested in maximizing their yields and profits. Although the metal concentrations in wastewater effluents are usually relatively low, long-term irrigation of land with such can eventually result in heavy metal accumulation in the soil.

Metal Mining and Milling Processes and Industrial Wastes

Mining and milling of metal ores coupled with industries have bequeathed many countries, the legacy of wide distribution of metal contaminants in soil. During mining, tailings (heavier and larger particles settled at the bottom of the flotation cell during mining) are directly discharged into natural depressions, including onsite wetlands resulting in elevated concentrations. Extensive Pb and Zn ore mining and smelting have resulted in contamination of soil that poses risk to human and ecological health. Many reclamation methods used for these sites are lengthy and expensive and may not restore soil productivity. Soil heavy metal environmental risk to humans is related to bioavailability. Assimilation pathways include the ingestion of plant material grown in (food chain), or the direct ingestion (oral bioavailability) of, contaminated soil.

Other materials are generated by a variety of industries such as textile, tanning, petrochemicals from accidental oil spills or utilization of petroleum-based products, pesticides, and pharmaceutical facilities and are highly variable in composition. Although some are disposed of on land, few have benefits to agriculture or forestry. In addition, many are potentially hazardous because of their contents of heavy metals (Cr, Pb, and Zn) or toxic organic compounds and are seldom, if ever, applied to land. Others are very low in plant nutrients or have no soil conditioning properties.

Air-borne Sources

Air-borne sources of metals include stack or duct emissions of air, gas, or vapor streams, and fugitive emissions such as dust from storage areas or waste piles. Metals from air-borne sources are generally released as particulates contained in the gas stream. Some metals such as As, Cd, and Pb can also volatilize during high-temperature processing. These metals will convert to oxides and condense as fine particulates unless a reducing atmosphere is maintained. Stack emissions can be distributed over a wide area by natural air currents until dry and/or wet precipitation mechanisms remove them from the gas stream. Fugitive emissions are often distributed over a much smaller area because emissions are made near the ground. In general, contaminant concentrations are lower in fugitive emissions compared to stack emissions. The type and concentration of metals emitted from both types of sources will depend on site-specific conditions. All solid particles in smoke from fires and in other emissions from factory chimneys are eventually deposited on land or sea; most forms of fossil fuels contain some heavy metals and this is, therefore, a form of contamination which has been continuing on a large scale since the industrial revolution began. For example, very high concentration of Cd, Pb, and Zn has been found in plants and soils adjacent to smelting works. Another major source of soil contamination is the aerial emission of Pb from the combustion of petrol containing tetraethyl lead; this contributes substantially to the content of Pb in soils in urban areas and in those adjacent to major roads. Zn and Cd may also be added to soils adjacent to roads, the sources being tyres, and lubricant oils.

Basic Soil Chemistry and Potential Risks of Heavy Metals

The most common heavy metals found at contaminated sites, in order of abundance are Pb, Cr, As, Zn, Cd, Cu, and Hg. Those metals are important since they are capable of decreasing crop production due to the risk of bioaccumulation and biomagnification in the food chain. There's also the risk of superficial and groundwater contamination. Knowledge of the basic chemistry, environmental, and associated health effects of these heavy metals is necessary in understanding their speciation, bioavailability, and remedial options. The fate and transport of a heavy metal in soil depends significantly on the chemical form and speciation of the metal. Once in the soil, heavy metals are adsorbed by initial fast reactions (minutes, hours), followed by slow adsorption reactions (days, years) and are, therefore, redistributed into different chemical forms with varying bioavailability, mobility, and toxicity. This distribution is believed to be controlled by reactions of heavy metals in soils such as (i) mineral precipitation and dissolution, (ii) ion exchange, adsorption, and desorption, (iii) aqueous complexation, (iv) biological immobilization and mobilization, and (v) plant uptake.

Lead

Lead is a metal belonging to group IV and period 6 of the periodic table with atomic number atomic mass 207, density $11.4\,g\,cm^{-3}$ melting point 327.4 °C, and boiling point 1725 °C. It is a naturally occurring, bluish-gray metal usually found as a mineral combined with other elements, such as sulphur (i.e., PbS, $PbSO_4$), or oxygen ($PbCO_3$), and ranges from 10 to $30\,mg\,kg^{-1}$ in the earth's crust. Typical mean Pb concentration for surface soils worldwide averages $32\,mg\,kg^{-1}$ and ranges from 10 to $67\,mg\,kg^{-1}$. Lead ranks fifth behind Fe, Cu, Al, and Zn in industrial production of metals. About half of the Pb used in the U.S. goes for the manufacture of Pb storage batteries. Other uses include solders, bearings, cable covers, ammunition, plumbing, pigments, and caulking. Metals commonly alloyed with Pb are antimony (Sb) (in storage batteries), calcium (Ca) and tin (Sn) (in maintenance-free storage batteries), silver (Ag) (for solder and anodes), strontium (Sr) and Sn (as anodes in electrowinning processes), tellurium (Te) (pipe and sheet in chemical installations and nuclear shielding), Sn (solders), and antimony, and Sn (sleeve bearings, printing, and high-detail castings).

Ionic lead, Pb(II), lead oxides and hydroxides, and lead-metal oxyanion complexes are the general forms of Pb that are released into the soil, groundwater, and surface waters. The most stable forms of lead are Pb(II) and lead-hydroxy complexes. Lead(II) is the most common and reactive form of Pb, forming mononuclear and polynuclear oxides and hydroxides. The predominant insoluble Pb compounds are lead phosphates, lead carbonates (form when the pH is above 6), and lead (hydr)oxides. Lead sulfide (PbS) is the most stable solid form within the soil matrix and forms under reducing conditions, when increased concentrations of sulfide are present. Under anaerobic conditions a volatile organolead (tetramethyl lead) can be formed due to microbial alkylation.

Lead(II) compounds are predominantly ionic (e.g., $Pb^{2+}\,SO_4^{2-}$), whereas Pb(IV) compounds tend to be covalent [e.g., tetraethyl lead, $Pb(C_2H_5)_4$]. Some Pb (IV) compounds, such as PbO are strong oxidants. Lead forms several basic salts, such as $Pb(OH)_2 \cdot 2P$-bCO which was once the most widely used white paint pigment and the source of considerable chronic lead poisoning to children who ate peeling white paint. Many compounds of Pb(II) and a few Pb(IV) compounds are useful. The two most common of these are lead dioxide and lead sulphate, which are participants in the reversible reaction that occurs during the charge and discharge of lead storage battery.

In addition to the inorganic compounds of lead, there are a number of organolead compounds such as tetraethyl lead. The toxicities and environmental effects of organolead compounds are particularly noteworthy because of the former widespread use and distribution of tetraethyllead as a gasoline additive. Although more than 1000 organolead compounds have been synthesized, those of commercial and toxicological importance are largely limited to the alkyl (methyl and ethyl) lead compounds and their salts (e.g., dimethyldiethyllead, trimethyllead chloride, and diethyllead dichloride).

Inhalation and ingestion are the two routes of exposure, and the effects from both are the same. Pb accumulates in the body organs (i.e., brain), which may lead to poisoning (plumbism) or even death. The gastrointestinal tract, kidneys, and central nervous system are also affected by the presence of lead. Children exposed to lead are at risk for impaired development, lower IQ, shortened attention span, hyperactivity, and mental deterioration, with children under the age of six being at a more substantial risk. Adults usually experience decreased reaction time, loss of memory, nausea, insomnia, anorexia, and weakness of the joints when exposed to lead. Lead is not an essential element. It is well known to be toxic and its effects have been more extensively reviewed than the effects of other trace metals. Lead can cause serious injury to the brain, nervous system, red blood cells, and kidneys. Exposure to lead can result in a wide range of biological effects depending on the level and duration of exposure. Various effects occur over a broad range of doses, with the developing young and infants being more sensitive than adults. Lead poisoning, which is so severe as to cause evident illness, is now very rare. Lead performs no known essential function in the human body, it can merely do harm after uptake from food, air, or water. Lead is a particularly dangerous chemical, as it can accumulate in individual organisms, but also in entire food chains.

The most serious source of exposure to soil lead is through direct ingestion (eating) of contaminated soil or dust. In general, plants do not absorb or accumulate lead. However, in soils testing high in lead, it is possible for some lead to be taken up. Studies have shown that lead does not readily accumulate in the fruiting parts of vegetable and fruit crops (e.g., corn, beans, squash, tomatoes, strawberries, and apples). Higher concentrations are more likely to be found in leafy vegetables (e.g., lettuce) and on the surface of root crops (e.g., carrots). Since plants do not take up large quantities of soil lead, the lead levels in soil considered safe for plants will be much higher than soil lead levels where eating of soil is a concern (pica). Generally, it has been considered safe to use garden produce grown in soils with total lead levels less than 300 ppm. The risk of lead poisoning through the food chain increases as the soil lead level rises above this concentration. Even at soil levels above 300 ppm, most of the risk is from lead contaminated soil or dust deposits on the plants rather than from uptake of lead by the plant.

Chromium

Chromium is a first-row d-block transition metal of group VIB in the periodic table with the following properties: atomic number atomic mass 24, density 7.19 g cm^{-3} melting point 1875 °C, and boiling point 2665 °C. It is one of the less common elements and does not occur naturally in elemental form, but only in compounds. Chromium is mined as a primary ore product in the form of the mineral chromite, $FeCr_2O_4$. Major sources of Cr-contamination include releases from electroplating processes and the disposal of Cr containing wastes. Chromium(VI) is the form of Cr commonly found at contaminated sites. Chromium can also occur in the +III oxidation state, depending on pH and redox conditions. Chromium(VI) is the dominant form of Cr in shallow aquifers where aerobic conditions exist. Chromium(VI) can be reduced to Cr(III) by soil organic matter, S^{2-} and Fe^{2+} ions under anaerobic conditions often encountered in deeper groundwater. Major Cr(VI) species include chromate (CrO_4^{2-}) and dichromate ($Cr_2O_7^{2-}$) which precipitate readily in the presence of metal cations (especially Ba^{2+}, Pb^{2+}, and Ag^+). Chromate and dichromate also adsorb on soil surfaces, especially iron and aluminum oxides. Chromium(III) is the dominant form of Cr at low pH (<4). Cr^{3+} forms solution complexes with NH OH$^-$, Cl$^-$, F$^-$, CN$^-$, SO_4^{2-}, and soluble organic ligands. Chromium(VI) is the more toxic form of chromium and is also more mobile. Chromium(III) mobility is decreased by adsorption to clays and oxide minerals below pH 5 and low solubility above pH 5 due to the formation of $Cr(OH)_3(s)$. Chromium mobility depends on sorption characteristics of the soil, including clay content, iron oxide content, and the amount of organic matter present. Chromium can be transported by surface runoff to surface waters in its soluble or precipitated form. Soluble and un-adsorbed chromium complexes can leach from soil into groundwater. The leachability of Cr(VI) increases as soil pH increases. Most of Cr released into natural waters is particle associated, however, and is ultimately deposited into the sediment. Chromium is associated with allergic dermatitis in humans.

Arsenic

Arsenic is a metalloid in group VA and period 4 of the periodic table that occurs in a wide variety of minerals, mainly as As_2O and can be recovered from processing of ores containing mostly Cu, Pb, Zn, Ag and Au. It is also present in ashes from coal combustion. Arsenic has the following properties: atomic number atomic mass density 5.72 g cm$^-$ melting point 817 °C, and boiling point 613 °C, and exhibits fairly complex chemistry and can be present in several oxidation states (−III, III, V). In aerobic environments, As (V) is dominant, usually in the form of arsenate (AsO_4^{3-}) in various protonation states: H_3AsO $H_2AsO_4^-$, $HAsO_4^{2-}$, and AsO_4^{3-}. Arsenate and other anionic forms of arsenic behave as chelates and can precipitate when metal cations are present.

Metal arsenate complexes are stable only under certain conditions. Arsenic (V) can also coprecipitate with or adsorb onto iron oxyhydroxides under acidic and moderately reducing conditions. Coprecipitates are immobile under these conditions, but arsenic mobility increases as pH increases. Under reducing conditions As(III) dominates, existing as arsenite (AsO_3^{3-}), and its protonated forms H_3AsO $H_2AsO_3^-$, and $HAsO_3^{2-}$. Arsenite can adsorb or coprecipitate with metal sulfides and has a high affinity for other sulfur compounds. Elemental arsenic and arsine, AsH may be present under extreme reducing conditions. Biotransformation (via methylation) of arsenic creates methylated derivatives of arsine, such as dimethyl arsine $HAs(CH_3)_2$ and trimethylarsine $As(CH_3)_3$ which are highly volatile. Since arsenic is often present in anionic form, it does not form complexes with simple anions such as Cl^- and SO_4^{2-}. Arsenic speciation also includes organometallic forms such as methylarsinic acid $(CH_3)AsO_2H_2$ and dimethylarsinic acid $(CH_3)_2AsO_2H$. Many As compounds adsorb strongly to soils and are therefore transported only over short distances in groundwater and surface water. Arsenic is associated with skin damage, increased risk of cancer, and problems with circulatory system.

Zinc

Zinc is a transition metal with the following characteristics: period group IIB, atomic number atomic mass 65, density $7.14\,g\,cm^{-3}$ melting point 419.5 °C, and boiling point 906 °C. Zinc occurs naturally in soil (about $70\,mg\,kg^{-1}$ in crustal rocks), but Zn concentrations are rising unnaturally, due to anthropogenic additions. Most Zn is added during industrial activities, such as mining, coal, and waste combustion and steel processing. Many foodstuffs contain certain concentrations of Zn. Drinking water also contains certain amounts of Zn, which may be higher when it is stored in metal tanks. Industrial sources or toxic waste sites may cause the concentrations of Zn in drinking water to reach levels that can cause health problems. Zinc is a trace element that is essential for human health. Zinc shortages can cause birth defects. The world's Zn production is still on the rise which means that more and more Zn ends up in the environment. Water is polluted with Zn, due to the presence of large quantities present in the wastewater of industrial plants. A consequence is that Zn-polluted sludge is continually being deposited by rivers on their banks. Zinc may also increase the acidity of waters. Some fish can accumulate Zn in their bodies, when they live in Zn-contaminated waterways. When Zn enters the bodies of these fish, it is able to biomagnify up the food chain. Water-soluble zinc that is located in soils can contaminate groundwater. Plants often have a Zn uptake that their systems cannot handle, due to the accumulation of Zn in soils. Finally, Zn can interrupt the activity in soils, as it negatively influences the activity of microorganisms and earthworms, thus retarding the breakdown of organic matter.

Cadmium

Cadmium is located at the end of the second row of transition elements with atomic number atomic weight 112, density $8.65\,g\,cm^{-3}$ melting point 320.9 °C, and boiling point 765 °C. Together with Hg and Pb, Cd is one of the big three heavy metal poisons and is not known for any essential biological function. In its compounds, Cd occurs as the divalent Cd(II) ion. Cadmium is directly below Zn in the periodic table and has a chemical similarity to that of Zn, an essential micronutrient for plants and animals. This may account in part for Cd's toxicity; because Zn being an essential trace element, its substitution by Cd may cause the malfunctioning of metabolic processes.

The most significant use of Cd is in Ni/Cd batteries, as rechargeable or secondary power sources exhibiting high output, long life, low maintenance, and high tolerance to physical and electrical stress. Cadmium coatings provide good corrosion resistance coating to vessels and other vehicles, particularly in high-stress environments such as marine and aerospace. Other uses of cadmium are as pigments, stabilizers for polyvinyl chloride (PVC), in alloys and electronic compounds. Cadmium is also present as an impurity in several products, including phosphate fertilizers, detergents and refined petroleum products. In addition, acid rain and the resulting acidification of soils and surface waters have increased the geochemical mobility of Cd, and as a result its surface-water concentrations tend to increase as lake water pH decreases. Cadmium is produced as an inevitable byproduct of Zn and occasionally lead refining. The application of agricultural inputs such as fertilizers, pesticides, and biosolids (sewage sludge), the disposal of industrial wastes or the deposition of atmospheric contaminants increases the total concentration of Cd in soils, and the bioavailability of this Cd determines whether plant Cd uptake occurs to a significant degree. Cadmium is very biopersistent but has few toxicological properties and, once absorbed by an organism, remains resident for many years.

Cadmium in the body is known to affect several enzymes. It is believed that the renal damage that results in proteinuria is the result of Cd adversely affecting enzymes responsible for reabsorption of proteins in kidney tubules. Cadmium also reduces the activity of delta-aminolevulinic acid synthetase, arylsulfatase, alcohol dehydrogenase, and lipoamide dehydrogenase, whereas it enhances the activity of delta-aminolevulinic acid dehydratase, pyruvate dehydrogenase, and pyruvate decarboxylase. The most spectacular and publicized occurrence of cadmium poisoning resulted from dietary intake of cadmium by people in the Jintsu River Valley, near Fuchu, Japan. The victims were afflicted by itai itai disease, which means ouch, ouch in Japanese. The symptoms are the result of painful osteomalacia (bone disease) combined with kidney malfunction. Cadmium poisoning in the Jintsu River Valley was attributed to irrigated rice contaminated from an upstream mine producing Pb, Zn, and Cd. The major threat to human health is chronic accumulation in the kidneys leading to kidney dysfunction. Food intake and tobacco smoking are the main routes by which Cd enters the body.

Copper

Copper is a transition metal which belongs to period 4 and group IB of the periodic table with atomic number atomic weight 63, density 8.96 g cm^{-3} melting point 1083 °C and boiling point 2595 °C. The metal's average density and concentrations in crustal rocks are 8.1 × 10^3 kg m^{-3} and 55 mg kg$^-$ respectively.

Copper is the third most used metal in the world. Copper is an essential micronutrient required in the growth of both plants and animals. In humans, it helps in the production of blood haemoglobin. In plants, Cu is especially important in seed production, disease resistance, and regulation of water. Copper is indeed essential, but in high doses it can cause anaemia, liver and kidney damage, and stomach and intestinal irritation. Copper normally occurs in drinking water from Cu pipes, as well as from additives designed to control algal growth. While Cu's interaction with the environment is complex, research shows that most Cu introduced into the environment is, or rapidly becomes, stable and results in a form which does not pose a risk to the environment. In fact, unlike some man-made materials, Cu is not magnified in the body or bioaccumulated in the food chain. In the soil, Cu strongly complexes to the organic implying that only a small fraction of copper will be found in solution as ionic copper, Cu(II). The solubility of Cu is drastically increased at pH 5. which is rather close to the ideal farmland pH of 6.0–6.5.

Copper and Zn are two important essential elements for plants, microorganisms, animals, and humans. The connection between soil and water contamination and metal uptake by plants is determined by many chemical and physical soil factors as well as the physiological properties of the crops. Soils contaminated with trace metals may pose both direct and indirect threats: direct, through negative effects of metals on crop growth and yield, and indirect, by entering the human food chain with a potentially negative impact on human health. Even a reduction of crop yield by a few percent could lead to a significant long-term loss in production and income. Some food importers are now specifying acceptable maximum contents of metals in food, which might limit the possibility for the farmers to export their contaminated crops.

Mercury

Mercury belongs to same group of the periodic table with Zn and Cd. It is the only liquid metal at stp. It has atomic number atomic weight 200, density 13.6 g cm^{-3} melting point −13.6 °C, and boiling point 357 °C and is usually recovered as a byproduct of ore processing. Release of Hg from coal combustion is a major source of Hg contamination. Releases from manometers at pressure-measuring stations along gas/oil pipelines also contribute to Hg contamination. After release to the environment, Hg usually exists in mercuric (Hg^{2+}), mercurous (Hg_2^{2+}), elemental (Hg^0), or alkylated form (methyl/ethyl mercury). The redox potential and pH of the system determine the stable forms of Hg that will be present. Mercurous and mercuric mercury are more stable under oxidizing conditions. When mildly reducing conditions exist, organic or inorganic Hg may be reduced to elemental Hg, which may then be converted to alkylated forms by biotic or abiotic processes. Mercury is most toxic in its alkylated forms which are soluble in water and volatile in air. Mercury(II) forms strong complexes with a variety of both inorganic and organic ligands, making it very soluble in oxidized aquatic systems. Sorption to soils, sediments, and humic materials is an important mechanism for the removal of Hg from solution. Sorption is pH dependent and increases as pH increases. Mercury may also be removed from solution by coprecipitation with sulphides. Under anaerobic conditions, both organic and inorganic forms of Hg may be converted to alkylated forms by microbial activity, such as by sulfur-reducing bacteria. Elemental mercury may also be formed under anaerobic conditions by demethylation of methyl mercury, or by reduction of Hg(II). Acidic conditions (pH < 4) also favor the formation of methyl mercury, whereas higher pH values favor precipitation of HgS(s). Mercury is associated with kidney damage.

Nickel

Nickel is a transition element with atomic number 28 and atomic weight 58.69. In low pH regions, the metal exists in the form of the nickelous ion, Ni(II). In neutral to slightly alkaline solutions, it precipitates as nickelous hydroxide, Ni(OH) which is a stable compound. This precipitate readily dissolves in acid solutions forming Ni(III) and in very alkaline conditions; it forms nickelite ion, HNiO that is soluble in water. In very oxidizing and alkaline conditions, nickel exists in form of the stable nickelo-nickelic oxide, Ni_3O that is soluble in acid solutions. Other nickel oxides such as nickelic oxide, Ni_2O and nickel peroxide, NiO are unstable in alkaline solutions and decompose by giving off oxygen. In acidic regions, however, these solids dissolve producing Ni^{2+}.

Nickel is an element that occurs in the environment only at very low levels and is essential in small doses, but it can be dangerous when the maximum tolerable amounts are exceeded. This can cause various kinds of cancer on different sites within the bodies of animals, mainly of those that live near refineries. The most common application of Ni is an ingredient of steel and other metal products. The major sources of nickel contamination in the soil are metal plating industries, combustion of fossil fuels, and nickel mining and electroplating. It is released into the air by power plants and trash incinerators and settles to the ground after undergoing precipitation reactions. It usually takes a long time for nickel to be removed from air. Nickel can also end up in surface water when it is a part of wastewater streams. The larger part of all Ni compounds that are released to the environment will adsorb to sediment or soil particles and become immobile as a result. In acidic soils, however, Ni becomes more mobile and often leaches down to the adjacent groundwater. Microorganisms can also suffer from growth decline due to the presence of Ni, but they usually develop resistance to Ni after a while. Nickel is not known to accumulate in plants or animals and as a result Ni has not been found to biomagnify up the food chain. For animals Ni is an essential foodstuff in small amounts. The primary source of mercury is the sulphide ore cinnabar.

Pesticides

A single teaspoon of healthy soil holds billions of soil microorganisms, including bacteria, fungi and other tiny life forms. These organisms have been sequestering carbon for hundreds of millions of years. They form symbiotic relationships with plant roots through mycorrhizal fungi. These networks help plants access nutrients like nitrogen and phosphorus from the soil in exchange for a steady flow of carbon in the form of carbohydrates the plant photosynthesizes from the air. The flow of carbon to the soil depends on this partnership between plant roots and soilmicroorganisms. But toxic pesticides can damage this microbial bridge.

Pesticides — a term that encompasses herbicides, insecticides and fungicides — are chemical compounds designed to kill, each with their own targets and mechanisms of action. As little as 0.1 percent of an applied pesticide interacts with its targeted weed or pest. The remainder contaminates the soil, air and water and can have significant nontarget effects throughout the ecosystem.

Pesticides can undercut regenerative agriculture goals by harming soil communities and altering critical biochemical processes in the soil.

Disrupting Soil Communities

Pesticides can cause significant changes in the composition, diversity and basic functioning of important soil microflora. This soil life is critical both to carbon sequestration and to a thriving, sustainable agriculture system. Some studies have found that pesticides decrease soil microbial biomass; others indicate that while total microbial biomass may not change, pesticides can reduce the abundance and diversity of soil organisms, damaging and altering important dynamics in the soil community. Some organisms can be suppressed while others may proliferate in the resulting vacant ecological niches, so organisms that were rare become abundant and vice versa. Research also shows that pesticides can impact larger fauna that help maintain the structure and fertility of the soil. For example, by impairing the reproductive capacity and survival of earthworms and springtails.

Increasing Bacterial Populations over Fungal Populations

Maintaining the right balance between bacterial and fungal populations in the soil is important for capturing more carbon. Pesticides can produce the opposite, increasing the proportion of bacterial to fungal populations in soil, according to several studies.

Altering Biochemical Processes

Soil microbes and plants make enzymes that catalyze biochemical transformation; these enzymes are the drivers of carbon and nutrient cycling. Research shows that some pesticides can interfere with enzyme production, inhibiting some while stimulating others, thus altering soil fertility, nutrient cycling and metabolism.

Hindering Nitrogen Fixation

Scientific evidence demonstrates that some pesticides can also hinder nitrogen fixation — another key regenerative agriculture aim — by inhibiting molecular communication between plants and rhizobia, the bacteria that fix nitrogen inside legume roots, and by diminishing root growth and reducing the number of root sites available for essential rhizobia.

No-till Farming and Glyphosate

Reducing soil tillage is an important principal of regenerative agriculture. However, advocates should be aware of the connection between no-till farming and glyphosate use as well as the latest science examining whether or not no-till agriculture increases overall soil carbon stocks.

While some farmers are succeeding at no-till farming with few chemical inputs, data indicates that the majority of notill farmers rely on herbicides such as glyphosate, the active ingredient in Roundup. In fact, 86 percent of No-Till Farmer readers said they planned to plant Roundup Ready corn in 2017, while 80 percent planned to plant Roundup Ready soybeans, and some 92 percent planned to use glyphosate for weed control.

Although the effects of glyphosate on soil are not fully understood, mounting evidence shows cause for concern. Studies have found that glyphosate damages the ecology of mycorrhizal fungi that enable the flow of carbon to the soil. Earthworms, so central to a healthy soil ecosystem, are also at risk from glyphosate exposure. One study found that the casting activity of earthworms at the soil surface nearly disappeared after three weeks of glyphosate application and that reproduction of earthworms dropped by half after three months.

Other potential negative effects of glyphosate on soil health include an increase in pathogenic microorganisms in the soil, impairment of respiration of soildwelling organisms and nutrient immobilization for plants and microorganisms.

The science is clear that routine use of glyphosate results in the development of resistant weeds. "Superweeds" now plague more than 60 million acres of U.S. farmland driving an intensification of herbicide use with new crops genetically engineered to withstand hazardous herbicides dicamba and 2,4-D. A 2018 survey conducted by No-Till Farmer found that 43 percent of growers surveyed planned to plant dicamba-resistant soybeans that season. The pesticide treadmill that often results from routine use of agricultural pesticides increases the toxic burden on soil life.

Meanwhile, the science on the benefits of no-till for soil carbon sequestration is not settled. Early studies found an increase in soil carbon in no-till systems, but that research focused only on the top foot of soil. Probing deeper into the soil profile, research suggests that no-till farming may redistribute soil carbon from deeper to shallower zones rather than increase the overall stock of soil carbon. This effect could cause more carbon releases from soil, rather than storing it deep in the ground where it is more stable — particularly as intermittent tillage may be important in no-till systems in some regions.

From Laboratory to Living Soil

The available scientific evidence likely underestimates pesticides' harm to soil health for a number of reasons. For one, we are just beginning to understand the factors that contribute to soil health and how the soil community functions. Additionally, research on pesticides and soil health is typically conducted in the laboratory rather than in the field, thus missing the complex, living nature of the soil biotic community. And most studies focus on acute impacts, failing to evaluate the chronic effects of pesticides on soil health.

Another significant shortcoming of the available research is that most studies focus on individual pesticides, whereas typical field use includes multiple pesticides at a time and over the course of a season. Research shows that mixtures of pesticide residues in the soil are the rule rather than the exception. The few studies that examine pesticide mixtures show that they can be more toxic than individual pesticides; mixtures can have synergistic or compounding effects. One study of two herbicides found that when each was applied alone, they exerted minor effects on the soil microbial community, but when combined, the impact was 10 times greater. Another study found that a mixture of glyphosate and diflufenican (used on soybeans) increased the toxic effects of both herbicides on soil biological activity as well as the persistence of each herbicide in the soil.

Protecting the Living Soil and All Life

Not only do pesticides pose a threat to the core aims of regenerative agriculture by harming the complex living community of the soil, mounting evidence shows that overuse of pesticides is decimating pollinators and other insects that are central to a sustainable food system. Scientists warn that we are in the midst of a "second silent spring" as populations of insects and birds suffer drastic declines. A recent global metaanalysis points to agricultural pesticides as a key driver of plummeting insect numbers. The researchers predict that over 40 percent of insect species may face extinction in coming decades, leading to widespread ecosystem collapse, if we don't change the way we farm. And the most comprehensive scientific assessment to date warns that loss of biodiversity is a global challenge on par with climate change.

Along with the environmental costs of agricultural pesticides, the human health costs are "catastrophic" according to a recent United Nations report. Yet, over one billion pounds of pesticides are used in the U.S. every year — and use of pesticides in the U.S. and globally is increasing.

Phasing out toxic pesticides must be a core part of what it means to be "regenerative" for life below ground as well as above and to ensure that regenerative agriculture doesn't become coopted or used as a cover for maintaining chemical-intensive industrial agriculture.

The good news is that regenerative agriculture practices like cover cropping and crop rotations can reduce farmers' need for toxic pesticides. A recent study in Nature reveals that most farmers could decrease pesticide use while maintaining or even improving their productivity. And decades of data debunk the myth that pesticides are necessary to feed a growing world population.

As our climate and biodiversity crises worsen, regenerative farming offers a crucial path to growing food in a way that nourishes both people and the planet. Innovative solutions should be guided by the best available science. And the science shows that eliminating or greatly reducing toxic pesticides is key to building healthy soils and ecosystems for a healthy planet.

Organic Pesticides

Pesticides are quite frequently used to -control several types of pests now-a-days. Pesticides may exert harmful effects to micro-organisms, as a result of which plant growth may beaffected. Pesticides which are not rapidly decomposed may create such problems. Accumu lation is residues of pesticides in higher concentrations are toxic. Pesticides persistence in soil and movement into water streams may also lead to their entry into foods and create health hazards. Pesticides particularly aromatic organic compounds are not degraded rapidly and therefore, have a long persistence time which can be seen in table.

Table: Persistence time for some selected pesticides.

Sl.No	Pesticide	Persistence time
1	BHC	11 yrs
2	DDT	10 yrs
3	2,4-D	2-8 weeks
4	Aldrin	9 yrs
5	Diuron	16 months
6	Atrazine	18 months
7	Siwazine	17 months

| 8 | Chlordane | 12 yrs |
| 9 | 2,3 6-Trichlorobenzene (TBA) | 2-5 yrs |

Mercury, cadmium and arsenic are common constituents of pesticides and all these heavy metals are toxic. At present DDT and a number of organochlorine compounds used as pesticides have been declared harmful and banned in U.S.A. and England. It is due to the persistence of their residues in soils for considerable time without losing their toxicity. This has led to higher concentration of these pesticides in vegetation, in animal flesh and milk. Eventually man has been affected. In view of their demerits, organochlorines have been replaced by organophosphate pesticides which are more toxic, but do not leave any residue. They do not pollute the soil. The rodenticides too add to soil pollution. A major method of checking this pesticidal pollution is to increase the organic matter content of the sol and choose such pesticides which are non-persistent and leave no harmful residue.

Organic Wastes

Organic wastes of various types cause pollution hazards. Domestic garbage, municipal sewage and industrial wastes when left in heaps or improperly disposed seriously affect health of human beings, plants and animals. Organic wastes contain borates, phosphates, deter- gents in large amounts. If untreated they will affect the vegetative growth of plants. The main organic contaminants are phenols and coal.

Asbestos, combustible materials, gases like methane, carbon dioxide, hydrogen sulphide, carbon monoxide, sulphur dioxide, petrol are also contaminants. The radioactive materials like uranium, thorium, strontium etc. also cause dangerous soil pollution. Fallout of strontium mostly remains on the soil and is concentrated in the sediments. Decontamination procedures may include continuous cropping and use of chelate amendments. Other liquids wastes like sewage, sewage sludge, etc. are also important sources of soil problems.

a. Sewage and sewage sludge:

Soil pollution is often caused by the uncontrolled disposal of sewage and other liquid wastes resulting from domestic uses of water, industrial wastes containing a variety of pollutants, agricultural effluents from animal husbandry and drainage of irrigation water and urban runoff. Irrigation with sewage water causes profound changes in the irrigated soils. Amongst various changes that are brought about in the soil as an outlet of sewage irrigation include physical changes like leaching, changes in humus content, and porosity etc., chemical changes like soil reaction, base exchange status, salinity, quantity and availability of nutrientslike nitrogen, potash, phosphorus, etc. Sewage sludges pollute the soil by accumulating the metals like lead, nickel, zinc, cadmium, etc. This may lead to the phytoxicity of plants.

b. Heavy metal pollutants:

Heavy metals are elements having a density greater than five in their elemental form. They mostly find specific absorption sites in the soil where they are retained very strongly either on the inorganic or organic colloids. They are widely distributed in the environment, soils, plants, animals and in their tissues. These are essential for plants and animals in trace amounts. Mainly urban and industrial aerosols, combustion of fuels, liquid and solid from animals and human beings, mining wastes, industrial and agricultural chemicals etc. are contributing heavy metal pollution. Heavy metals are present in all uncontaminated soils as the result of weathering from their parent materials. Concentration of heavy metals in soils and plants is given in table.

Table: Heavy metal concentration in the hithosphere, soils and plants (Ug/gm dry matter).

Sl.No	Heavy metal	Hithosphere	Soil range	Plants
1	Cadmium (Cd)	0.2	0.01-0.7	0.2-0.8
2	Cobalt (Co)	40	1-40	0.05-0.5
3	Chromium (Cr)	200	5-3000	0.2-1.0
4	Copper (Cu)	70	2-100	4-15
5	Iron (Fe)	50,000	7000-5,50,000	140
6	Mercury (Hg)	0.5	0.01-0.3	0.015
7	Manganese (Mn)	1000	100-4000	15-100
8	Molybdenum (Mo)	2.3	0.2-5	1-10
9	Nickel (Ni)	100	10-1000	1
10	Lead (Pb)	16	2-200	0.1-10
11	Tin (Sn)	40	2-100	0.3
12	Zinc (Zn)	80	10-300	8-100

In agricultural soils, however, the concentration of one or more of these elements may be significantly increased in several ways, like through applications of chemicals, sewage sludge, farm slurries, etc. Increased doses of fertilizers, pesticides or agricultural chemicals, over a period, add heavy metals to soils which may contaminate them. Certain phosphatic fertilizers frequently contain trace amounts of cadmium which may accumulate in these soils. Likewise, some fertilizers when applied to soils, they add certain heavy metals which are given in table.

Table: Heavy metal content of fertilizers (ug/gm).

Sl.No	Fertilizer	Co	Cr	Cu	Mn	Mo	Ni	Pb	Zn
1	Nitrochalk	-	-	22	24	-	2	-	15
2	Calcium	0.1	Traces	Traces	Traces	-	-	-	1
3	Nitrate	-	-	To 10	To 5	-	-	-	-

4	Ammonium sulphate	<5	<5	0.800	0.80	<0.05 to	0.22	<5	Traces to 200
5	Super phosphate 0.02-13 0-1000	Traces to 1000	Traces to 2842	Traces to 35	Traces to 32	Traces to 92	70-3000		
6	Potassium chloride								
7	Potassium sulphate	<5	<5	0-300 to 80	Traces to . 33	0.09	<5	<50	<50

Table: Heavy metal contents in sludges (ppm).

Sl.No	Heavy metal	Range (ppm)
1	Cadmium	< 60-1500
2	Cobalt	2-260
3	Chromium	40-8800
4	Copper	200-8000
5	Iron	6000-62,000
6	Manganese	150-2500
7	Molybdenum	2-30
8	Nickel	20-5300
9	Lead	120-3000
10	Zinc	700-49,000

The fate of heavy metals in soil will be controlled by physical and biological processes acting within the soil. Metal ions enter the soil solution from these various forms of combination in different rates they may either remain in solution or pass into the drainage water or be taken up by plants growing on the soil or be retained by the soil in sparingly soluble or insoluble forms. The organic matter of these soil have great affinity to heavy metals cations which form stable complexes thereby leading to reduced nutrient content.

Herbicides

When a herbicide is used to control weeds, sometimes a majority of the compound ends up in the environment, whether it is in the soil, water, atmosphere or in the products harvested. Due to the widespread use of these chemicals over the years, there has been an accumulation of these residues in the environment, which is causing alarming contaminations in the ecosystems and negative damages to the biota. To Bolognesi and Merlo, the widespread use of herbicides has drawn the attention of researchers concerned with the risks that they can promote on the environment and human health, since they are chemicals considered contaminants commonly present in hydric resources and soils. Herbicides represent a high toxicity to target species but it can be also toxic, at different levels, to non-target species, such as human beings. Herbicides can cause deleterious effects on organisms and human health, both by their direct and indirect action. Among the biological effects of these chemicals, it can be cited genetic damages, diverse physiological alterations and even death of the organisms exposed. Some herbicides, when at low concentrations, cannot cause immediate detectable effects in the organisms, but, in long term can reduce their lifespan longevity. Herbicides can affect the organisms in different ways. As with other pesticides, the accumulation rate of these chemicals on biota depends on the type of the associated food chain, besides the physicochemical characteristics (chemical stability, solubility, photo-decomposition, sorption in the soil) of the herbicide. Thus, despite the existence of several toxicological studies carried out with herbicides, in different organisms, to quantify the impacts of these pollutants and know their mechanisms of action. There is a great need to expand even more the knowledge about the effects of different herbicides in aquatic and terrestrial ecosystems. Data obtained from in situ, ex situ, in vivo and in vitro tests, derived from experiments of simulation, occupational exposure or environmental contaminations, need to enhance so that it is possible to obtain even more consistent information about the action of these compounds.

When herbicides are applied in agricultural areas they can have different destinations, since being degraded by microorganisms or by non-biological means or even be transported by water, to areas distant from the application site. The organisms can be then exposed to a great number of these xenobiotics as well as their metabolites.

During the last two decades, several studies have been completed to predict the behaviour of pesticides in the soil. Despite the numerous efforts to assess the effects of herbicides in the soil, there are conflicting data in the literature on the subject, where some studies show that the residues of pesticides can be sources of carbon and energy to microorganisms, and then are degraded and assimilated by them, while other reports affirm that pesticides produce deleterious effects to the organisms and biochemical and enzymatic processes in the soil. In general, the application of pesticides, and here it is also included herbicides, made long term, can cause a disturbance in the biochemical balance of the soil, which can reduce its fertility and productivity.

Once in the soil, herbicides can suffer alteration in their structure and composition, due to the action of physical, chemical and biological processes. This action on the herbicides is the one that will determine their activity and persistence in the soil. Some molecules, when incorporated into the soil, are reduced by volatilization and photo-decomposition. Once in the soil, herbicides can suffer the action of microorganisms, which, added to the high humidity and high temperature, can have their decomposition favoured. If they are not absorbed by plants, they can become strongly adsorbed on the organic matter present in the colloidal fraction of the soil, be carried by rainwater and/or irrigation and even be leachate, thus reaching surface or groundwater.

The prediction of the availability of herbicides to plants has two purposes: i. ensure that the herbicide reaches the roots in concentrations high enough to control weeds, without compromising the agricultural productivity; ii. predict if the compound is mobile in the soil to estimate how much of the herbicide can be leachate from the roots zone to groundwater.

The contamination of aquatic environments by herbicides has been characterized as a major world concern. This aquatic contamination is due to the use of these products in the control of aquatic plants, leachate and runoff of agricultural areas.It is a growing public concern about the amount of herbicides that have been introduced into the environment by leachate and runoff, not to mention that the contaminations of the aquatic environments generally occur by a mixture of these compounds and not by isolated substances.

Guzzella et al. did a survey on the presence of herbicides in groundwater in a highly cultivated region of northern Italy. The researchers monitored for two years the presence of 5 active ingredients and 17 metabolites resulting from these compounds.

Toccalino et al. carried out a study to verify the potential of chemical mixtures existing in samples of groundwater used for public supply. In these samples, the most common organic contaminants were herbicides, disinfection by-products and solvents. The combined concentrations of the contaminants can be a potential concern for more than half of the samples, even though the water destined to public supply pass through treatments to reduce contaminations and meet the legislations, it can still contain mixtures at worrying concentrations.

Saka evaluated the toxicity of three herbicides (simetryn, mefenacet and thiobencarb) commonly used in rice planting in Japan, on the test organism Silurana tropicalis (tadpoles). The authors observed that the three herbicides, particularly thiobencarb, are toxic for tadpoles (LD50 test), even for concentrations found in waters where the rice is cultivated. In a similar study carried out by Liu et al., it was observed that the effect of the herbicide butachlor (most used herbicide in rice planting in Taiwan and Southeast Asia) on the organism Fejervarya limnocharis (alpine cricket frog) exposed to concentrations used in the field. In this study no effect on the growth of tadpoles of F. limnocharis was observed, but there was a negative action on survival, development and time of metamorphosis. The herbicide butachlor can cause serious impacts on anurans that reproduce in rice fields, but this impact varies from species to species.

In a study, it was observed that the herbicide atrazine has a genotoxic and mutagenic effect on the species Oreochromis niloticus (Nile tilapia). In this study, the authors observed that the herbicide can interfere in the genetic material of the organisms exposed, even at doses considered residual, which led the authors to suggest that residual doses of atrazine, resulting from leaching of soils of crops near water bodies, can interfere in a negative form in the stability of aquatic ecosystems.

Bouilly et al. studied the impact of the herbicide diuron on Crassostrea gigas (Pacific oyster) and observed that the herbicide can cause irreversible damages to the genetic material of the organism studied. Moreover, due to the persistence of diuron in environments adjacent to its application site and that it is preferably used in spring, the pollution caused by its use causes negative impact in the aquatic organisms during the breeding season.

In general, when herbicides contaminate the aquatic ecosystem, they can cause deleterious effects on the organisms of this system. Thus, organisms that live in regions impacted by these substances, whose breeding period coincides with the application period of the herbicides, can suffer serious risks of development and survival of their offspring.

ladik et al. evaluated the presence of two herbicides (chloroacetamide and triazine), as well as their by-products, in drinking water samples of the Midwest region of the United States. The authors detected the presence of neutral chloroacetamide degradates in median concentrations (1 to 50 ng/L) of the water samples. Furthermore, they found that neither the original chloroacetamide herbicides nor their degradation products were efficiently removed by conventional water treatment processes (coagulation/flocculation, filtration, chlorination). According to Bannink, about 40% of the drinking water from Netherlands is derived from surface water. The Dutch water companies are facing problems with the water quality due to contamination by herbicides used to eliminate ruderal plants. These data serve as alerts for the presence of herbicides and their degradation products in drinking water, pointing out the need for the development of new treatment systems that could be more efficient to eliminate this class of contaminants.

Organic herbicides, when in aquatic ecosystems, can be distributed in several compartments depending on their solubility in water. These compartments include water, aquatic organisms, suspended sediment and bottom sediment. The more hydrophilic the organic pesticide, the more it is transported to the aqueous phase, and the more hydrophobic a pesticide is, the more it will be associated to the organic carbon of the suspended and bottom sediment. The sorption of the herbicides in sediments in suspension can reduce the degradation rate of the herbicides in water, and the movement of the sediment in suspension can transport the pesticides from one place to another, entering into the tissue of organisms or settling on the bottom.

A study evaluated the contamination of three matrices (water, sediment and bivalve molluscs) collected in rivers influenced by crops of sugar cane in São Paulo State-Brazil. In this study, the authors observed that the highest concentrations of residues of the herbicide ametrin were present in the sediment, showing the persistence of this compound in the sediments of rivers and its potential to mobilize between the compartments of the aquatic system, such as water and biota.

When the herbicides are dispersed in the water or sediments in suspension of the rivers, they can end up in other ecosystems such as estuaries. Duke et al., when studying the effect of herbicides on mangroves of the Mackay region, found out that diuron, and even other herbicides, are potentially responsible for the mangrove dieback. The consequences for this death would be the impoverishment of the quality of the coastal water with an increase of the turbidity, nutrients and sediment deposition, problems in the fixation of seedlings and consequent erosion of the estuaries.

Soil Pollution through Transport Activities

Construction of transport infrastructure can have significant environmental impacts if not undertaken with care. Transportation activities support increasing mobility demands for passengers and freight notably in urban areas; on the other hand, they are associated with high levels of environmental externalities. This has reached a point where transportation is a dominant source of emission of most pollutants and their multiple impacts on the environment. These impacts fall within three categories: Direct impacts (the immediate consequence of transport activities on the environment), Indirect impacts (the secondary effects of transport activities on environmental systems, often of higher consequence than direct impacts, but the involved relationships are often difficult to establish) and Cumulative impacts (the additive, multiplicative or synergetic consequences of transport activities).

The relationships between transport and the environment are multidimensional. Some aspects are unknown and some new findings may lead to drastic changes in environmental policies. The 1990s were characterized by a realization of global environmental issues, epitomized by the growing concerns between anthropogenic effect and climate change. Transportation also became an important dimension of the concept of sustainability, which is expected to become the prime focus of transport activities in the coming decades, ranging from vehicle emissions to green supply chain management practices.

Transport activities have resulted in growing levels of motorization and congestion. The most important impacts of transport on the environment relate to climate change, air quality, noise, water quality, biodiversity, land take, and soil quality. The environmental impact of transportation on soil consists of soil erosion and soil contamination. The removal of earth's surface for highway construction or lessening surface grades for port and airport developments have led to important loss of fertile and productive soils. Soil contamination can occur through the use of toxic materials by the transport industry. Fuel and oil spills from motor vehicles are washed on road sides and enter the soil. Chemicals used for the preservation of railroad ties may enter into the soil. Hazardous materials and heavy metals have been found in areas contiguous to railroads, ports and airports.

Pedosphere is the surface cover of the earth's crust which is under the soil formatting processes' activity. The European Soil Card issued in 1972 by the European Council says that soil nourishes plants and, indirectly, animals and people. At the same time there should be mentioned that the soil soaks in the decayed remains of plants and animals exchanging them into the alimentary components for plants or the soil-formatting substances. Soil as environment component always has a direct or indirect influence on people's health and life. The pollutants from the soils can penetrate to the human organism through the skin, lungs (air pollution) and oesophagus (food originated from plants and animals) or indirect contact. After many years of trials and errors, people have realized that the environment cannot be divided into separate elements and learnt to treat it as a whole system. Also, the pedosphere cannot be described as a separate unit but as a part of a larger system.

Roadside Soil Pollution by Chemicals

In non-urban region there is not any issue of atmospheric pollution by CO, CO_2 and NO_x. Thus, this issue is not examined in rural and periurban roads. On the contrary, as the particles originating from the pavement, its maintenance or the traffic enter the ground transferred by the water other pollutants are also transferred by the wind and are spread in various distances polluting the soil. As to the total soil pollution (transferred by the wind and the water) elements requiring special caution are the heavy metals (lead, zinc, cadmium, nickel), the sodium chloride, hydrocarbons and dusts.

Chemical extractants, including the complexing agent DTPA (diethylenetriaminepentaacetic acid), have been used trying to evaluate heavy metal availability in different kinds of soil and grass. Lindsay and Norvell working with neutral and calcareous soils deficient in trace metals proposed DTPA soil test. Then, the use has been extended to other types of soils even the contaminated by different sources. Total (soil and grass) and DTPA-extractable (soil) contents of Cd, Cu, Fe, Mn, Pb and Zn could be determined by Atomic Absorption Spectroscopy. The metal contents in road soils and grasses confirm the effect of traffic as a source of pollution. The most remarkable influence has been shown in soil, since it accumulates metals on a relatively long-term period.

The main processes by which vehicles spread heavy metals (Pb, Zn, Cu, Cd, Ni) into the environment are combustion processes, the wear of cars (tires, brakes, engine), leaking of oil and corrosion. Certain components of automotive engines, chasis and piping contain manganese and copper, while chromium and nickel (also coming from combustion of lubricating oils) are used in chrome plating. Lead is released in combustion of leaded petrol, zinc is derived from tire dust, copper is derived from brake abrasion and corrosion of radiators, and the other heavy metals have mixed origins. Heavy metals are also released due to weathering of road surface asphalt and corrosion of crash barriers and road signs. Studies on heavy metals have shown that the concentrations in the soil are closely related to vehicle's traffic in road's vicinity, as well as to the distance from the road. The quality guidelines for plant and soil heavy metal concentrations developed in various countries indicate wide variations. The natural heavy metal content of soil is strongly related to the composition of the parent rock. The mean lead concentration for surface soil on the world scale is estimated as 25 parts per million (ppm) with the upper limit at 70 ppm. Mean total zinc and copper contents in surface soils of different countries range from 17 to 125 ppm and 6 to 60 ppm, respectively. Soils throughout the world contain nickel within a broad range (1 to about 200 ppm). The background cadmium levels in soils do not exceed 0.5 ppm, while the calculated mean of nickel for world soils is 20 ppm. Because of the severe adverse environmental and/or ecological and health effects of heavy metals, there have been many studies on heavy metal contamination in soils along major roads.

Road traffic and maintenance induce continual heavy metal pollution in roadside soil and runoff water. A portion of these pollutants could be dispersed into the atmosphere or deposited onto soils as a result of wind dispersion. The atmospheric deposition and quality of roadside soil have been investigated alongside two major rural highways. Metal deposition decreased rapidly and seemed to reach the background level at a distance of less than 25 m. The deposition of zinc was found to be the most significant (galvanized crash barriers constitute an additional source of zinc pollution), followed by Pb (high bioavailability of lead in polluted soils was shown.) and Cu. Pollutant concentrations in soils decrease rapidly with distance from the roadway. The pollution was concentrated within a 5 m zone, while the lead content exceeded the limit value for contaminated soil at a distance of 0.50 m from the road. The lead concentration in the ground in various distances from the pavement depends on the size of particles, 75% of which has a diameter smaller than 5 microns; hence, they could be carried by the wind in large distances. However, except the cases where vigorous winds or local atmospheric inversions prevail, it seems that significant concentrations are presented in distances between 5 and meters from the pavement and that heavy metal concentrations in the ground beyond 10 meters are negligible.

Heavy metals' and other pollutants concentrations are carried from the ground to the plants, especially to the plant generation following the first pollution (cultivation cycle). The concentration will increase and pass to the food chain. Forages' or meadow grass pollution by which beef are fed in the course of several years may accumulate in their meat, reaching a high concentration degree which may be harmful and dangerous for human's eating this meat health. Plants absorb metals mainly via the roots and their respiratory organs (the leaves) where the pollutants are seated. It is estimated that 50 to 60% of the seated lead may be removed with a simple wash. With the increasing use of unleaded petrol, the lead levels tend to decrease regularly and therefore another tracer should be identified in order to assess the road transport contamination, like platinum of catalytic silencers or some additives in unleaded petrol like MTBE (methyltertiobuthylether), ETBE (ethyltertiobuthylether), TAME (methyltertioamylether).

Cadmium (Cd), another heavy metal, is cumulative poison posing problems for humans and preferably accumulates in the kidneys. The Pb/Cd ratio in plants, taking it from the ground, is reduced as the concentration in them increases. That is, plants prefer the cadium concentrations. The ratio is even smaller for humans. So it is possible in extreme cases to cause poisoning (the deathly dose to humans is about 400 mg).

Hydrocarbons impact is of secondary importance, in the case of time impact associated with automobile's movement. Hydrocarbons' time pollution modifies soil's natural properties (increases the imperviousness with time) and blocks plant's respiration since it seats on the leaves. Dusts have a similar behaviour, as they also seat on the leaves blocking the air alternation. There is no doubt that most children, especially toddlers, ingest a significant amount of dust and dirt from their hands. The most obvious routes are thumb and finger sucking and via sweets and food. The mass of street dust ingested daily will vary widely from child to child, and although direct measurements of this mass are scarcely practicable, some estimate of an average daily intake can be made.

Sodium chloride, which is the most used ice melting material, has an impact on soil and on plants. In a distance up to three meters from the roads carriageway its concentrations are due to the flow of salt water from the pavement as well as to splashings; between thre and eight meters are due mainli to splashings. Finally, the increase of salination beyond the eight meters becomes negligible. Sodium chloride acts in the ground increasing its structural instability, reducing its permeability and diminishing the stability of other cations, especially calcium in the complex humus-clay. Experiments on plants in road central islands have shown that salt has a direct effect with leaves scorching and on the metabolism with the absorption via the roots. This last impact is mainly associated with salt's accumulation on the leaves introducing hydraulic disorders and decrease of animal ions in plants, such as calcium and potassium. These phenomenons are especially noticeable in evergreen plants.

The relatively smallest threats are: salinity and the soil's structure and porosity destruction – because of different reasons. Salinity – because of really good mitigation measures necessary to be used when constructing a motorway or express road (like drainage systems, ditches etc.). At the same time soil's structure destruction by the building machines is limited in time and takes place at limited, not huge area along the road. Moreover, this impact is of short duration and its effects are, in their majority, convertible.

Mobilized heavy metals become readily available to plants and they enter into the food chain. They can also migrate to groundwater. Specific soil's properties, mainly its pH, Eh, cation exchange capacity, amount of organic matter in soil, amount of clay minerals and amount of iron, manganese and aluminum oxides, control the rates of heavy metal migration in the profiles. Clay minerals, Fe-, Mn-, Al- oxides and organic matter are the most important groups for the sorption of heavy metals. Heavy metals can be also incorporated in the structures of carbonates, phosphates and sulfides.

Soil samples from the Shen-Ha (Shenyang-Harbin) Highway, Northeast China, at and 320 m distances had been used in order to investigate the effect of heavy metals of highway origin on soil nematode guilds. The contents of total and available Pb, Cu, Zn varied significantly with the different distances from the highway. Pb was the main pollutant in the soils in the vicinity of Shen-Ha Highway. The soil-band 20 to 40 m away from the highway was the most polluted area. Thirty six genera of soil nematodes belonging to 23 families were identified, which may act as a prominent indicator to heavy metal pollution of highway origin.

Boundary, alerting and critical values of heavy metals in soils (in mg/kg of dry soil or in ppm) are given by country's legislations. For example, according to Slovenian legislation (Official Gazette, 68/96), the corresponding values are: For Pb 100 and for Zn 300 and for Cu 100 and for Ni 70 and and for Cd 2 and 12 respectively.

Protective Measures against Soil Pollution

The protective measures may be distinguished in preventive and curative. As a protective measure, olericulture cultivations should be avoided within a distance less than 10 metres from the road when this is in embankment or in the same grade with the natural soil. The use of plants growing within a distance smaller than 10 metres from a heavily trafficked road, should strictly be avoided for animal feeding. This holds true for pastures or meadows, but for forage from slope production or road's central islands, also. Curative measures could be described as follows:

- In the case of exceptionally sensitive zone and for its protection the creation of a coppice type (silvicolous) hedge, with a height equal to the slope's height, when the road is on an embankment, with a minimum height of 2 metres.

- For the creation of the above mentioned hedges in urban or suburban zones, the planting of plants bearing up the pollution is preferred, avoiding plants with continuous leafage as well as plants with smooth and spoil leaves, more sensitive in the accumulation of dust.

- In areas where considerable use of chemical solvents is made, the planting of plants that bear up the salt should be provided, aiming at the maintenance of good planting (draining, good quality of natural ground, insemination, suitable planting season).Whenever possible, calcium chloride ($CaCl_2$) must be used instead of sodium chloride which is more toxic for the plants and the soil.

For the erosion and sediment control, a site-specific erosion and sediment control plan has to be developed in order to minimize the impacts of runoff waters on construction activities. A number of provisions to lessen the environmental impacts of road construction are specified in an erosion and sediment control plan, including measures to ensure that exposed working surfaces are kept to a minimum, silt fences and sediment traps are optimally placed to prevent sediment from reaching drainage systems, vehicles are washed when leaving a construction site to remove excess mud, and temporary exit/entry roads to construction sites are provided with a coarse rock surface to prevent the transfer of soil offsite where it will be washed into nearby drainage channels.

On road embankment slopes, slopes of all cut, fill etc., shrubs and grass will be planted. On sections with high filling and deep cutting the side slopes will be graded and covered with bushes and grass, etc., adopting suitable bioengineering techniques, the suitability to be decided by the Engineer at site. Along sections abutting water bodies stone pitching needs to be carried out for slopes between 1 vertical: 4 horizontal to 1 vertical: 2 horizontal. At the outfall of each culvert, erosion prevention measures, such as grass scales, rock riprap, rock mattresses, cut off wall and downstream silt screens/walls will be undertaken, as provided in the design: The work shall consist of measures as per design, or as directed by the Engineer to control soil erosion, sedimentation and water pollution, through the use of berms, dikes, sediment basins, fibre mats, mulches, grasses, slope drains and other devices. All temporary sedimentation, pollution control works and maintenance thereof will be deemed as incidental to the earth work or other items of work.

Transport is essential to the economic and social development, by providing access to jobs, housing, goods and services and providing for the mobility needs of people to improve quality of life.

Appraising total soil resistance to transport threats is necessary to take into account soil's resistance to every separate threat as well as connections between its properties. The change of ground water conditions is considered as the relatively highest threat. The subsoil water level lowering is usually followed by negative effects appearing gradually and on the other hand soil flooding results in gley formatting that destroys soil irreversibly. The accumulation of heavy metals, especially cadmium compounds, is a large threat also. These pollutants are of significant ecological/environmental concern because they are not biodegradable and have long half-lives in the soil, thus predicating far reaching effects on biological systems including soil micro-organisms and other soil biota. Actually, there are no methods for their removal from the soils. Moreover, there are no methods of avoiding their emissions. Nevertheless, zones of elevated concentrations of heavy metals compounds are not very wide (~50 m from the roadway's edge). Therefore, it is only safe to use grass for feeding animals or to eat vegetables grown at distances more than 50 meters from the edge of highways.

The concentration of soil pollution is dangerous both for the agricultural products, directly taken by humans, and for fodder ending up again to man via animals meat. Pollutants, particularly lead and zinc, accumulate in the roadside soils and are absorbed by invertebrate macrofauna and vegetation. The general decrease in concentrations of heavy metals with distance from the road, with depth in the soil profile, and the general increase in concentrations with traffic density indicates their relation to traffic. With the increasing use of unleaded petrol, the lead levels tend to decrease regularly and therefore another tracer should be identified in order to assess the road transport contamination. Lead seems to be the most adapted tracer of highway contamination. The plant bioaccumulation results enable researchers to understand the metal bioavailability in the environment.

Due to large number of motor vehicles on Greek roads, and as a consequence of commercial and industrial activities, considerable amounts of some heavy metals are likely to be emitted regularly as long as the nearby sources remain active.

References

- Soil-pollution, environmental-chemistry: toppr.com
- Causes-and-effects-of-soil-pollution: conserve-energy-future.com, Retrieved 19 April, 2019
- Soil-pollution: byjus.com, Retrieved 23 June, 2019
- Soil-pollution-by-transportation-projects-and-operations- 313598720: researchgate.net, Retrieved 15 March, 2019

2

Causes of Soil Pollution

Soil pollution is caused by various natural and man-made factors which include acid rain, overgrazing, urbanisation, landfill and illegal dumping, mining, use of agrochemicals and petrochemicals, etc. This chapter has been carefully written to provide an easy understanding of these different causes of soil pollution.

Natural Causes

Natural factors due to which soil degradation occurs include volcanic eruptions, alterations in rainfall patterns, earthquakes, geographical changes, air pollution and melting of glaciers.

At times, natural accumulation of chemicals results in soil pollution while sometimes natural processes also raise pollutant toxicity and soil pollution level. Soil has complex properties that can contain both chemicals and natural conditions that can interact with one another to produce pollution. Following are some natural processes that cause soil pollution:

- Due to the imbalance between the atmospheric deposition and leakage with rain water, the natural accumulation of compounds takes place in the soil (accumulation of perchlorate in dry environment).

- Natural production of perchlorate takes place in the soil in some environmental conditions (in the presence of metal objects and energy generated by the storm).

- It is also caused due to sewer lines leakage in the subsurface of soil.

Acid Rain

Soil is the basis of wealth upon which all land-based life depends. The damage that occurs to ecosystems from acidic deposition is dependent on the buffering ability of that ecosystem. This buffering ability is dependent on a number of factors, the two major ones being soil chemistry and the inherent ecosystem sensitivity to acidification. Indirect damage to ecosystems is largely caused by changes in the soil chemistry. Increasing soil acidity can affect micro-organisms which break down organic matter into nutrient form for plants to take up. Increasing soil acidity also allows aluminium (a common constituent of soil minerals) to come into solution. In its free organic form, aluminium is toxic to plant roots and can lock up phosphate, thereby reducing the concentrations of this important plant nutrient.

Effect of Acid Rain on Soil and Underlying Bedrock

Soils containing calcium and limestone are more able to neutralise sulphuric and nitric acid depositions than a thin layer of sand or gravel with a granite base.

If the soil is rich in limestone or if the underlying bedrock is either composed of limestone or marble, then the acid rain may be neutralised. This is because limestone and marble are more alkaline (basic) and produce a higher pH when dissolved in water. The higher pH of these materials dissolved in water offsets or buffers the acidity of the rainwater producing a more neutral pH.

Acid Sensitive Areas

In regions where the soil is not rich in limestone or if the bedrock is not composed of limestone or marble, then no neutralising effect takes place, and the acid rainwater accumulates in the bodies of water in the area. This applies to much of the north-eastern United States where the bedrock is typically composed of granite. Granite has no neutralising effect on acid rainwater. Therefore over time more and more acid precipitation accumulates in lakes and ponds.

The water bodies most susceptible to change due to acid precipitation are those whose catchments have shallow soil cover and poorly weathering bedrock, for example granite and quartzite. These soil types are characterised by the absence of carbonates that could neutralise acidity. The run-off water from such areas is less buffered than from areas such as limestone catchments, with an adequate level of carbonate. Such catchments and waters are termed acid-sensitive (poorly buffered), and can suffer serious ecological damage due to artificially acidified precipitation from air masses downwind of major emissions.

Effects of Soil on Vegetation

When acid rain falls, it can affect forests as well as lakes and rivers. To grow, trees need healthy soil to develop in. Acid rain is absorbed into the soil making it virtually impossible for these trees to survive. As a result of this, trees are more susceptible to viruses, fungi and insect pests.

Long-term changes in the chemistry of some sensitive soils may have already occurred as a result of acid rain. As acid rain moves through the soils, it can strip away vital plant nutrients through chemical reactions, thus posing a potential threat to future forest productivity.

Poisonous metals such as aluminium, cadmium and mercury, are leached from soils through reacting with acids. This happens because these metals are bound to the soil under normal conditions, but the added dissolving action of hydrogen ions causes rocks and small-bound soil particles to break down.

Plant life in areas where acid rain is common may grow more slowly or die as a result of soil acidification. In the Green Mountains of Vermont and the White Mountains of New Hampshire in the United States 50% of the red spruce have died in the past 25 years. There has also been noted a reduced amount of growth in existing trees as measured by the size of growth rings of the trees in these areas.

These effects occur because acid rain leaches many of the existing soil nutrients from the soil. The number of micro-organisms present in the soil also decreases as the soil becomes more acidic. This further depletes the amount of nutrients available to plant life because the micro-organisms play an important role in releasing nutrients from decaying organic material. In addition, the roots of plants trying to survive in acidic soil may be damaged directly by the acids present. Finally, if the plant life does not die from these effects, then it may be weakened enough so that it will be more susceptible to disease or other harsh environmental influences like cold winters or high winds.

Critical Loads

Environmental response to pollutants depends on many factors. Some regions cope with acidification better than others, having larger 'critical loads'. Critical load refers to the greatest assault that an ecological system can withstand before showing measurable degradation.

Scientists determine critical load by examining rock and soil type, land use and rainfall. If soil is fertile with a pH greater than 4.5, and rainfall is relatively low, the critical load will be high. The terrain can withstand moderately large additions of acidity without undue suffering. Conversely, in low pH soils, acidification mobilises toxic aluminium ions. If coniferous forests predominate, or if land is devoted to rough grazing, the result is a low critical load. Even minor acid deposition is undesirable.

Man-made Causes

Industrial Waste

Industrial activity has been the biggest contributor to the problem in the last century, especially since the amount of mining and manufacturing has increased. Most industries are dependent on extracting minerals from the Earth. Whether it is iron ore or coal, the by-products are contaminated and they are not disposed of in a manner that can be considered safe. As a result, the industrial waste lingers in the soil surface for a long time and makes it unsuitable for use.

The increasing industrialization has led to the pollution of soil through the discharge of effluents by the industrial units. Each kind of soil has its own individuality. The distinctive feature of this individuality is the soil profile, which consists of series of layers different from one of percolation of the waste water discharged into the land and the subsequent washing down of the pollutants to the successive horizons. The effluents discharged by the industrial units onto land contain many toxic chemicals, mineral acids, bases etc. which over a period of time get deposited in the soil due to their retention and adsorption on the soil particles. The mineral constituents present in trace amounts in the discharged effluent favor the growth of some algal, fungal and bacterial colonies which in turn change the texture of the soil. The micronutrients discharged into the soil through effluents reduce the porosity of the soil resulting in poor yields. Also some of the deposited chemicals may be taken up by the plants/ crops growing in such contaminated soils. Organic effluent with high concentration of biodegradable organic matter discharged into the soil attract the saprophytic soil and air micro flora and thus could proliferate resulting in poor yields or fungal diseases in many cases. Thus there is a need to monitor the soils where waste water is applied for irrigation/ plantation purposes.

Agrochemicals

Agrochemicals are used in agricultural setting in an effort to ensure an abundant food supply. Many important benefits are achieved by the use of agrochemicals. These are largely associated with increased yields of plant and animal crops, and less spoilage during storage. These benefits are substantial. In combination with genetically improved varieties of crop species, agrochemicals have made important contributions to the successes of the "green revolution." This has helped to increase the food supply for the rapidly increasing population of humans on Earth. However, the use of certain agrochemicals has also been associated with some important environmental and ecological damages. Extensive application of external agricultural inputs to agricultural production systems leads to soil quality degradation. Organic (carbon-based) pollutants that impact soil quality include pesticides. Pesticides, which are very persistent in soil, slowly break down and result in source of contamination.

Soil acts as filter, buffer and degradation potentials with respect to storage of pollutant with the help of soil organic carbon, but it is recognized that the soil is a potential pathway of pesticide transport to contaminate water, air, plants, food and ultimately to human via, runoff and sub-surface drainage; interflow and leaching; and the transfer of mineral nutrients and pesticides from soil into the plants and animals that constitute the human food chain. The capacity of the soil to filter, buffer, degrade, immobilize, and detoxify pesticides is a function or quality of the soil. Soil quality also encompasses the impacts that soil use and management can have on water and air quality, and on human and animal health. Inappropriate use of chemical fertilisers and pesticides, amongst common farming practices, can contribute significantly to the soil degradation process. There is evidence that prolonged use of heavy doses of fertilisers can result in soils becoming more acidic that has serious implications in terms of long term productivity of soils. Inappropriate, viz. imbalanced or excessive, use of fertilisers is a major cause of pollution of ground waters or surface water bodies resulting from inefficient use of applied nutrients. Many of the chemicals used in pesticides are persistent soil contaminants, whose impact may endure for decades and adversely affect soil conservation. Pesticides enter the soil via spray drift during foliage treatment, wash-off from treated foliage, release from granulates or from treated seeds in soil. The presence and bio-availability of pesticides in soil can adversely impact human and animal health, and beneficial plants and soil organisms. These are detrimental to living organisms in the soil, vital to soil health and productivity. Pesticides can move off-site contaminating surface and groundwater and possibly causing adverse impacts on aquatic ecosystems.

Pesticides in soils continue to be studied more than any other environmental contaminant, because they are used widely to control pests that affect agricultural crops and pests in the home, yards, and gardens. Though beneficial, they contaminate soil eco-system and pose threat to the balance equilibrium among various groups of microorganisms and components in soil. There is also increasing interest in their transformation products (TPs), because they can be present at higher levels in soil than the parent pesticide itself. Most of the insecticides applied on the crops eventually end as accumulation in the soil. There are many ways in which a insecticide reaches the soil and are more risky in the soil as a result of their residues which may be comprises of many substances including any specified derivatives such as degradation products, metabolites and congeners that are considered to be of toxicological. Several studies have been made to detect the contaminated soil in which different kinds of pesticides have been found.

Those that were once released into air or water will end up in soils, with the exception of those that are deposited at the bottom of oceans. Among organic pollutants some are referred to as 'POPs,' or persistent organic pollutants, which do not break down quickly in the environment and persist for a longer period thereby resulting to various hazardous consequences. Continuous and excessive use of pesticide compounds has led to the contamination of several ecosystems in different parts of the world.

Environmental risk due to soil pollution is of particular importance for agricultural areas, as extensive reliance on agrochemicals for agricultural productions have immensely resulted in the accumulation of various heavy metals in the soil leading to serious consequences. Thus pollution of heavy metals poses a threat to a country's food production. Some fertilizers and pesticides are known to contain various levels of heavy metals, including Cd and Cu. Therefore, continuous and heavy application of these agrochemicals and other soil amendments can potentially exacerbate the accumulation of heavy metals in agricultural soils over time.

Mining

Mining can cause physical disturbances to the landscape, creating eyesores such as waste-rock piles and open pits. Such disturbances may contribute to the decline of wildlife and plant species in an area. In addition, it is possible that many of the premining surface features cannot be replaced after mining ceases. Mine subsidence (ground movements of the earth's surface due to the collapse of overlying strata into voids created by underground mining) can cause damage to buildings and roads. Between 1980 and 1985, nearly five hundred subsidence collapse features attributed to abandoned underground metal mines were identified in the vicinity of Galena, Kansas, where the mining of lead ores took place from 1850 to 1970. The entire area was reclaimed in 1994 and 1995.

Modern Agriculture

Industrial agriculture is a form of modern farming that refers to the industrialized production of livestock, poultry, fish, and crops. The methods of industrial agriculture are techno scientific, economic, and political. They include innovation in agricultural machinery and farming. Modern methods of agriculture have resulted in use of fertilizers and pesticides to increase the yield of the crops.

Most of them are synthetic and chemicals-based. They are collectively called agro-chemicals. When a nucleus emits radiation, it is said to decay. Rays from radioactive isotopes are dangerous to living things but are also useful in variety of ways. Exposure to large amounts of radiation is harmful to health.

The Greenhouse effect is the rise in temperature that the Earth experiences because certain gases in the atmosphere referred to as greenhouse gases, trap energy from the Sun.

- Destruction of wild life leads to loss of valuable materials.

- Disappearance of any link in a food chain upsets nature's balance and creates problems.

- New improved varieties of crops and other useful animals are derived from their wild relatives by genetic modification.

Degradation of Soil

Land degradation refers to a decline in the overall quality of soil, water or vegetation condition commonly caused by human activities. This leads to soil erosion, rising water tables, the expression of salinity, mass movement by gravity of soil or rock, stream bank instability and a process that results in declining water quality. Land degradation is a global problem, largely related to agricultural use.

The major causes include:

- Land clearance, such as clear cutting and deforestation.

- Agricultural depletion of soil nutrients through poor farming practices.

- Livestock including overgrazing.

- Irrigation and over drafting.

- Urban sprawl and commercial development.

- Land pollution including industrial waste.

- Vehicle off-roading.

- Quarrying of stone, sand, ore and minerals.

Impacts of Overgrazing

The huge population of livestock needs to be fed and the grazing lands or pasture areas are not adequate. Very often we find that the livestock grazing on a particular piece of grass-land or pasture surpass the carrying capacity. Following are the impacts of overgrazing.

Land Degradation

Overgrazing removes the vegetal cover over the soil and the exposed soil gets compacted due to which the operative soil depth declines. So the roots cannot go much deep into the soil and adequate soil moisture is not available.

Organic recycling also declines in the ecosystem because not enough detritus or litter remains on the soil to be decomposed. The humus, content of the soil decreases and overgrazing leads to organically poor, dry, compacted soil.

Due to trampling by cattle the soil loses infiltration capacity, which reduces percolation of water into the soil and as a result of this more water gets lost from the ecosystem along with surface run off. Thus overgrazing leads to multiple actions resulting in loss of soil structure, hydraulic conductivity and soil fertility.

Soil Erosion

Due to overgrazing by cattle, the cover of vegetation almost gets removed from the land. The soil becomes exposed and gets eroded by the action of strong wind, rainfall etc. the grass roots are very good binders of soil. When the grasses are removed, the soil becomes loose and susceptible to the action of wind and water.

Loss of Useful Species

Overgrazing adversely affects the composition of plant population and their regeneration capacity. The original grassland consists of good quality grasses and herbs with high nutritive value.

When the livestock graze upon them heavily, even the root stocks which carry the reserve food or regeneration get destroyed. Now some other species appear in their place. These secondary species are hardier and are less nutritive in nature. Some livestock keep on overgrazing these species also.

Peterochemicals

The petroleum is mainly composed of various hydrocarbon complex mixtures. It can be divided into saturated hydrocarbon, aromatic hydrocarbon and non-hydrocarbon compounds by chromatography. The saturated hydrocarbon molecular structure consists of carbon-carbon bond and carbonhydrogen bond, which is easy to be degraded. In addition, its boiling point is relatively low, so the saturated hydrocarbon can gradually disappear from the soil in the way of photosynthesis and volatilization. However, the molecular structure of aromatic hydrocarbon is quite complex. The complex benzene ring and its higher boiling point greatly increase the difficulties of being removed from the soil. The polycyclic aromatic hydrocarbon is the typical POPs that widely exists in various environmental systems, and the 16 kinds of PAHs have been included in the list of priority control pollutants by EPA of the United States and European Community, namely naphthalene, acenaphthene, acenaphthylene, fluorene, phenanthrene, anthracene, fluoranthene, pyrene, benzo[a]anthracene, benzo[a]pyrene, benzo[b]fluoranthene, benzo[k]fluoranthene, diphenyl[a,h]anthracene, benzene [ghi]perylene and indene and [1,2,3-cd] pyrene.

Among the three kinds of hydrocarbon compounds, non-hydrocarbon compounds have the maximum carbon number. They are insoluble in water, and the fusion and boiling point are quite high, so they are most difficult to be removed from the soil, resulting in obvious environmental toxicity and mutagenicity. Therefore, the natural soil residual poison composition is mainly polycyclic aromatic hydrocarbons and nonhydrocarbon compounds.

Besides oil, the sewage in oil and gas fields can also lead to soil pollution. At present, most domestic oil fields have entered into the middle and later periods of oil production, water content of the crude oil reaches to 70%~80%, some even up to 90%. Large amounts of oily sewage will produce after oil/water separation. If untreated, they could lead to serious soil and water pollution. In terms of its composition, the oilfield sewage contains oil, various salts, organic matter, inorganic matter and some microbes etc. The salinity is higher, generally between 10^3 ~14×10^4 mg/L, the main salts are Na_2SO_4, $NaHCO_3$, $MgCl_2$ and NaCl. This kind of waste water not only causes soil salinization, but also destroys the soil environment quality.

Table: The composition of oilfield sewage.

type	primary materials
crude oil	oil content 1000-2000mg/L, oil spill, dispersed oil, emulsified oil and dissolved oil

inorganic salts	$Ca^{2+}, Mg^{2+}, K^+, Na^+, Cl^-, HCO_3^-, CO^{2-}, SO_4^{2-}$
organics	aliphatic hydrocarbon, aromatic hydrocarbon, phenols, organic sulfide, aliphatic acid, polymers
inorganic matter	H_2S, FeS, clay particles, silt and fine sand
microorganism	sulfate reducing bacteria, saprophytic bacteria and iron bacteria

The Harm of Oil-polluted Soil

The harm of oil-polluted soil mainly includes the following aspects: Firstly, because of the small density, higher viscosity and lower emulsifying ability of petroleum, it is easy to be absorbed in soil surface, affecting the permeability and porosity of soil; petroleum is rich in carbon and a small amount of nitrogen compounds, so it can change the composition and structure of soil organic matter and impact the C/N, C/P, salinity, pH, EH and conductivity of soil. The heavy metals (nickel and vanadium) in oil mixtures and high concentrations of salt in oilfield output water can also damage the soil environment.

Secondly, microorganisms in natural environment are quite abundant in healthy and clean soil. In normal situation the microorganisms which can resist the oil pollution stress are not developed, while in contaminated soil, in order to adapt to this kind of environment, they can produce certain enzyme system and gradually form a dominant population with symbiotic or synergy effect. A number of studies have shown that the hydrocarbon pollution can change the microbial population, the composition of the community structure and the enzyme system in soil, given priority to the inhibitory action.

Thirdly, it can impede the normal growth of crops such as reduce the germination rate and fertility and decline the resistance to pests and diseases. In addition, the oil compounds could react with inorganic nitrogen and phosphorus, limiting the nitrification and removal of phosphoric acid, so the effective nitrogen and phosphorus in the soil would decrease and the absorption of crops will be affected.

Moreover, the polycyclic aromatic hydrocarbons in petroleum chemicals have carcinogenic, mutagenic, teratogenic and other toxic effects. It can enter into the bodies of people and animals through breathing, skin contact and diet, degrading the normal function of livers and kidney etc, therefore causing great threat to human's health.

At last, the oil pollutants in the soil not only impact the pedosphere, but also the atmosphere and water sphere. To be specific, the low boiling point and light weight hydrocarbons can enter into the atmosphere by evaporation easily; then through runoff and infiltration into the surface water and osmosis into the groundwater system; and finally through the food chain enter into the human's bodies.

Sewage Discharge

Soil pollution is often caused by the uncontrolled disposal of sewage and other liquid wastes resulting from domestic uses of water, industrial wastes containing a variety of pollutants, agricultural effluents from animal husbandry and drainage of irrigation water and urban runoff. Irrigation with sewage water causes profound changes in the irrigated soils. Amongst various changes that are brought about in the soil as an outlet of sewage irrigation include physical changes like leaching, changes in humus content, and porosity etc., chemical changes like soil reaction, base exchange status, salinity, quantity and availability of nutrients like nitrogen, potash, phosphorus, etc. Sewage sludges pollute the soil by accumulating the metals like lead, nickel, zinc, cadmium, etc. This may lead to the phytoxicity of plants.

Landfill and Illegal Dumping

Landfills directly render the soil and land where it is located unusable. It also destroys the adjacent soil and land area because the toxic chemicals spread over the surrounding soil with time. The upper layer of the soil is damaged, distorting soil fertility and activity and affecting plant life. Industrial and electronic wastes in the landfills destroy the quality of the soil and land thereby upsetting the land ecosystems.

Urbanization

Globally urbanization occurs rapidly and now land surface are being sealed increasingly all over the world. Soil has versatile functions and in urban area its ecological service, especially the ability to buffer and purify pollutants, is very much needed. However, due to various intensive human activities, soils in urban areas are often subject to fundamental changes even serious degradation. Their degradation is essentially a process of a sacrifice by providing ecological services, such as sink of all kinds of pollutants, at the cost of their own quality. Therefore, as a result, they often come across a variety of environmental problems.

Physical degradation, such as compaction, destruction of structure, may reduce the ability of urban soils in infiltrating water and storing capacity, thus causes a higher runoff and pollutant load to the receiving water bodies. Soil compaction is also believed to worsen city heat island effect by reflecting more radiation.

Enrichment of various waste materials associated with human activities, including nutrient elements, heavy metals, and organic pollution are the major problems of urban soil environment. The main features of urban soil contamination are characterized firstly by strong accumulation of so-called 'urban elements' such as Cu, Zn, Pb and Hg, but less of other heavy metals; and secondly by spatial isolation, which means the contamination is not spatially continuous. During the process of urban development, heavy metal contamination of urban soils happens not only nowadays, but also did in the past, especially when primitive mining and metal processing prevailed.

Furthermore, urban soils are often polluted by organic pollutants, especially polycyclic aromatic hydrocarbons (PAHs), with roadsides and industrial sites being the major vulnerable areas. PAHs therein have normally 2-6 rings and predominantly 2-4 rings, strongly suggesting their pyrogenic background, including traffic emission, industrial activities and coal burning.

So far, studies on urban soil contamination and other related environmental problems have concentrated mainly on three aspects, firstly source, status and diffusion patterns of urban soil contamination; secondly impact and risk assessment of soil contamination on environment, ecology and biological health; thirdly better use and management of urban soils.

Electronic Waste

Soil can be contaminated in two primary ways from e-waste: (a) through direct contact with contaminants from e-waste or the byproducts of e-waste recycling and disposal; or (b) indirectly through irrigation from contaminated water.

When e-waste is improperly disposed in regular landfills or illegally dumped, both heavy metals (lead, arsenic, cadmium, and others) and flame retardants in e-waste can leach directly from the e-waste into the soil, causing contamination of underlying groundwater or contaminating crops that may be planted in that soil now or in the future.

When e-waste is not recycled properly as is the case in areas of the world where re-cycling practices for e-waste are not regulated or are informally monitored, soil can become directly contaminated by (a) effluent or waste products from leaching practices which extract precious metals and other valuable materials from e-waste; (b) coarse particles and bottom ash generated from dismantling, shredding, or burning of e-waste; and (c) leaching of heavy metals not recovered during recycling into underlying soil during disposal. Practices used to extract precious metals from e-waste such as mercury amalgamation or cyanide leaching can release additional toxic substances to the soil. Dismantling can also release large, coarse particles into the air, which due to their size and weight, quickly re-deposit to the ground and subsequently contaminate soil. Shredding or burning of e-waste produces ash which can be heavily contaminated by both heavy metals and flame retardants (polybrominated diphenyl ethers or PB-DEs) that leach into underlying soil. By similar processes, heavy metals left over from incomplete recycling can also contaminate underlying soil. How and how much soil is contaminated depends on a wide range of factors including temperature, pH, soil type, climate, and soil composition. Much of this soil contamination is persistent and these pollutants remain in the soil for a long time, some evolving into even more toxic species than in their original form. Soil is also indirectly impacted by e-waste recycling through contact with contaminated water.

References

- Causes-and-effects-of-soil-pollution: conserve-energy-future.com, Retrieved 19 June, 2019

- Mining, Li-Na: pollutionissues.com, Retrieved 14 May, 2019

- Soil-pollution-impacts-of-modern-agriculture-and-degradation-on-soil- 27421: yourarticlelibrary.com, Retrieved 20 June, 2019

- Natural-resources/impacts-of-overgrazing-land-degradation-soil-erosion-and-loss-of-useful-species- 30033: yourarticlelibrary.com, Retrieved 15 April, 2019

- Soil-contamination-risk-assessment-and-remediation, environmental-risk-assessment-of-soil-contamination: intechopen.com, Retrieved 17 March, 2019

- Causes-effects-solutions-of-landfills: conserve-energy-future.com, Retrieved 14 February, 2019

3

Soil Pollution Monitoring and Assessment

Soil pollution monitoring and assessment refers to the prevention, management, risk assessment and mitigation of the toxic agents of soil that adversely affect the land, air, water and living organisms. The topics elaborated in this chapter will help in gaining a better perspective about soil pollution monitoring and assessment.

Soil Contamination Monitoring

Soil contamination by naturally occurring and anthropogenic organic and inorganic chemicals is a serious human and environmental health problem in many industrialized and nonindustrialized nations. There is a wide range of types of soil contamination, and an equally wide range of methods and approaches to soil monitoring. Practical considerations such as how the data will be used, the data's required accuracy and precision, and the amount of money, staff, and instrumentation available for the analysis also play a part in the selection of appropriate soil contamination monitoring methods.

Several approaches to soil contamination monitoring include chemical, geophysical, and biological techniques. Chemical techniques are used to measure specific organic, inorganic, or radioactive contaminants in the soil using instruments such as a gas chromatograph, atomic absorption spectrometer, or mass spectrometer. Geophysical techniques examine changes in physical properties of the soil and the contaminants to address large areas of soil contamination.

They may not require any disturbance to the soil, but may not be useful for identifying each contaminant. Biological techniques use organisms as indicators of soil contamination, or byproducts of contaminant biodegradation processes to monitor or predict changes in soil contaminant concentrations over time.

Current developments in soil contamination monitoring include increased efficiency of soil contaminant extraction processes that improve contaminant recovery, development of laboratory instrumentation with enhanced detection limits or ease of use, and development of alternative techniques for soil contamination monitoring such as isotopic signatures or immunoassays.

In addition, on-site analyses allow monitoring of soil without removing it from the site using portable and hand-held meters, and field kits. Some are research-based techniques that may become standard for soil contamination monitoring. At this time, additional development of innovative techniques is warranted that produces cost-effective, robust, easily used and sensitive monitoring techniques for organic, inorganic, and radioactive contaminants in soil.

Soil contamination by organic and inorganic contaminants has been recognized as an important problem in many areas of the industrialized nations. In addition, naturally occurring contaminants from radiological Earth sources and human and animal wastes (nutrients and pathogenic bacteria) all impact soil and sediments.

In many countries including the United States, land application of hazardous and radioactive wastes is used because it is economical relative to other types of waste disposal.

Industrial, military, and municipal waste disposal is often on the land surface or subsurface, that is, buried in both shallow and deep soils. Land applied contaminants filter through the soils and may impact ground water.

Waste disposal over several decades by application to land in shallow pits or ponds has become a problem in many countries that are now struggling with the task of remediating or cleaning up these areas. Of special concern are areas that pose a direct threat to human and environmental health.

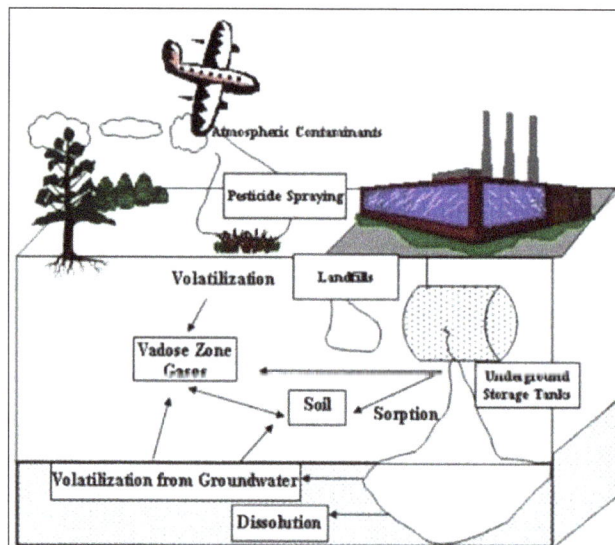

Soil contamination comes from multiple sources and is impacted by processes such as sorption to soil particles and volatilization into the vadose zone.

Soils can become contaminated with a wide range of pollutants from various sources other than land disposal of wastes. Contaminants may be applied directly to the soil, as is the case with pesticides. Alternatively, chemicals in soils can occur as a result of air pollutants that fall out as wet or dry deposition and settle on aquatic or land surfaces.

An example is the contamination of aquatic sediments from the deposition of hydrophobic chemicals emitted from hazardous waste incinerators. These pollutants fall out of the air onto lakes and are eventually trapped on the aquatic sediments where they can reside for many years.

Chemical pollutants in soils range in their properties from hydrophobic organic contaminants that are strongly associated with the soil, to more water soluble organic contaminants that are transported long distances and are partitioned primarily in the aqueous phase, to radioactive metals that have chemical characteristics like metals in addition to their radioactive properties. Because soil contamination ranges from metals, to complex organo-metallic compounds, to large and small molecular weight organic contaminants, soil monitoring presents many challenges and requires that varied techniques be used. Typical examples of soil contaminants occurring ubiquitously in the environment include hydrophobic polycyclic aromatic hydrocarbons (PAHs) which sorb readily to soils; chromium, a typical metallic contaminant in sediments associated with commercial harbors; tri-butyl tin, a complex organometallic compound used to prevent biofouling on ship hulls; and methyl tertiary butyl ether, a highly soluble gasoline additive which has been found to contaminate groundwater and sediments in several areas of the USA, particularly California.

Soil monitoring techniques are not only varied because of the range in chemical properties of the contaminants themselves, but also because the reason for which the monitoring is being done is varied. Some soil monitoring is used primarily as a screening test to estimate relative concentrations of contaminants or groups of contaminants. Some monitoring is required by local, state, or federal regulations and certified analytical laboratories are required to carry out the analyses. Other monitoring falls in between these two, and may be useful for monitoring the progress of natural attenuation, the natural cleaning up of contaminated soils due to chemical, biological, and physical processes in the environment. Each of these types of concerns requires a different level of analytical sophistication in terms of detection level, separation, and identification of specific chemical contaminants in soils. In addition, issues of cost, reliability of the measurement, and ease of the analysis are important in determining which analyses and for what purpose the analyses are used.

The main difficulty in soil monitoring arises from the nature of the soil matrix. Pollutants in water or in air generally are more easily measured than those associated with soil. This is due in large measure to the interaction of the contaminants with the soil particles themselves. Strong chemical and physical forces may act to bind the contaminants to the soil particles. Thus, if the monitoring technique requires that the chemicals be extracted or removed from the soil prior to analysis, the efficiency of the extraction process becomes crucial to the overall success of the analysis. A second problem is access to the contaminated soils. Land-applied contaminants migrate downward with time and become less accessible. Similarly, contaminants may be applied directly to deeper areas, and as depth of contaminated soils increases, monitoring techniques also may change.

The best techniques, generally speaking, are those that are nonobtrusive, inexpensive, and relatively easy to carry out using field sampling instruments. However, usually the reliability of the information and the difficulty of the analysis may be correlated, that is, the more reliable and sophisticated the analysis, the more difficult, time consuming and expensive it is. A big challenge in the field of soil monitoring is to provide relatively reliable soil monitoring methods that are easily carried out with minimal personnel training required and which use field-hardy, inexpensive instrumentation.

Certain techniques for soil monitoring have been used for several decades. These often are extensions of analytical chemistry techniques which have been adapted for soil analyses, or designed specifically for soil monitoring in situ, meaning in the field at the site of contamination. For example, traditional soil contamination monitoring would include the collection of the soil sample which is returned to the laboratory to measure inorganic and organic contaminants. Extraction of the contaminant from the soil is necessary. The extraction is followed by analysis by analytical methods such as gas chromatography, mass spectrometry, atomic adsorption spectrophotometry, fluorescence spectroscopy, and infrared nuclear magnetic resonance (NMR) spectroscopy.

Monitoring soil in the field without the benefit of laboratory analysis may be less expensive but adds uncertainty concerning the identification of specific chemicals. Field soil monitoring is carried out using specially designed instruments for nonobtrusive sampling in which soils do not always have to be removed prior to analysis. Alternatively, laboratory instruments are modified to be carried into the field for use after soils have been collected. Field test kits and field instruments are usually smaller, more portable and more resistant to field conditions of wear and tear, travel, and other forms of physical abuse than laboratory instruments. Examples of in situ soil monitoring techniques include chemical analyses using modified laboratory instruments including a portable gas chromatograph or field kit that includes all reagents and a portable spectrometer for colorimetric analyses; physical analyses of the contaminated areas using electrical conductivity; and biological analyses that include an assessment of highly sensitive species as indicators of pollution, or biomarkers.

The main objective of soil monitoring is to prevent and mitigate contamination by substances with the potential to exert an adverse effect on the soil itself, and on air, water and organisms that may contact the soil. Soil monitoring, within the approvals program, is directed primarily to the assessment of contaminants that have been released to the soil surface. Thus, subsurface facilities are generally not the reason for soil monitoring, but may be the reason for ground water monitoring. However, where soil contamination is known or suspected to originate from subsurface sources such as underground tanks or pipes, an assessment will be required. Where the above considerations indicate soil monitoring is required as a condition of an Approval, the proponent is required to carry out the following, as specified in the Soil Monitoring Directive:

- Prepare a soil monitoring proposal;

- Execute the approved soil monitoring plan;

- Interpret and report the results of the soil monitoring;

- Prepare and execute a soil management plan where indicated by the results of soil monitoring. This guideline provides a background for the soil monitoring program and a description of soil management program requirements.

Legislative Background

The soil monitoring program mostly developed under the Environmental Protection in support of the following principles:

- Development must be sustainable, meaning that the use of resources and the environment today must not impair prospects for their use by future generations;

- The environmental impact of development must be prevented or mitigated;

- Polluters should bear the responsibility of paying for the costs of their actions;

- Remediation costs should be incorporated into financial planning so that adequate funds are available for site remediation and planners can know the true costs and benefits of source reduction programs. Recognizing that under the environmental protection is a shared responsibility, it follows that both the approval holder and the Department must have a means to assess environmental performance with respect to the above principles and requirements.

Soil Quality Standards

Environmental Protection expects that approval holders will manage their operations to prevent substance releases to soil. Substance releases to soil do occur, however, and contam- inants are often present above background concentrations at industrial facilities. In view of this, Environmental Protection should have soil quality standards to guide assessment and remediation of soil contamination. Facilities that are currently uncontaminated have the opportunity to maintain conditions that allow unrestricted land-use. For these facilities, the minimum standards will be determined by the Tier I criteria or equivalent objectives. Older facilities, however, were often operated under different standards and environmental management practices than are currently acceptable.

Properties of Soil Contaminants

Table: Physical-chemical characteristics of contaminants that impact soil contamination monitoring.

Characteristic	Common abbreviation	Units	Environmental relevance
Molecular mass	MW	Atomic mass units	Mass of a contaminant.
Solubility	S	mg L^{-1} or g m^{-3}	Tendency of contaminant to dissolve in a liquid.
Density	D	g cm-3	Mass of a unit volume of contaminant.
Vapor pressure	v.p.	Pa	Tendency of contaminant to exist in the air phase.
Sorption coefficient	K_d	Dimensionless	Partitioning of contaminant between sediment and water.
Organic carbon partition coefficient	K_{oc}	Dimensionless	Partitioning of contaminant between sediment and water, corrected for organic carbon content of soil.
Octanol-water partition coefficient	K_{ow}	Dimensionless	Partitioning of contaminant between lipids and water; Estimate of hydrophobicity.
Henry's law constant	K_H	Pa-m^3 mole^{-1}	Partitioning of contaminant between air and water.
Radioactivity	None	Bq	Decay of a radionuclide; disintegrations per second.

The most important physicochemical properties and commonly used coefficients are defined in table. Knowing the physicochemical properties of a contaminant aids in determining which soil monitoring technique is appropriate. Solubility is one of the most readily available characteristics of contaminants and is defined as the concentration of a contaminant in equilibrium in a saturated solution at a given temperature. Solubility(mg L^{-1}) is usually a function of the contaminant's molecular weight and density. Solubility information will provide an indication of the contaminant's ability to remain in the aqueous phase and not sorb onto soil particles.

Sorption is a process by which a contaminant is chemically or physically bound to the soil particle. In many cases, the greater the strength of the chemical bonds, the greater the extraction process required to break the bonds and release the contaminant for analysis. Sorption in natural soils may be a function of the amount of organic carbon or other sorptive materials naturally present in the soil. For this reason coefficients have been derived to express not only the amount of a contaminant sorbed to soil relative to that remaining in the liquid phase, often termed K_d, but also the amount of a contaminant sorbed to soil corrected for the amount of organic carbon present in the soil relative to the amount of contaminant remaining in the liquid phase, K_{oc}. Figure shows a typical soil coring with lighter colored soil particles and decreasing organic carbon content with depth below land surface. The top layers of the soil will have a greater capacity to sorb hydrophobic contaminants than the deeper, sandier, less organic, carbon-rich soil. This sorption impacts the ease of extraction of the contaminant and therefore the ease of soil monitoring. Another coefficient, the octanol water partition coefficient (K_{ow}) is a second indirect indicator of the probability of a contaminant to sorb to soil and is a measure of hydrophobicity. It was derived to mimic partitioning between water and lipid materials such as those in organisms as an indicator of bioaccumulation of contaminants.

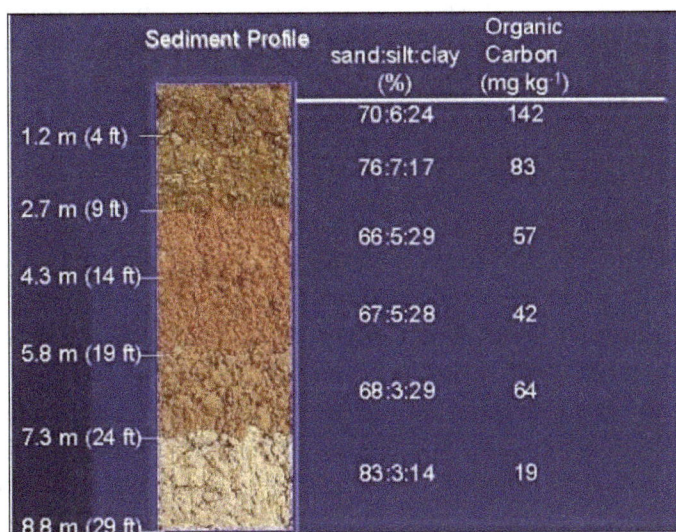

Sediment Profile	sand:silt:clay (%)	Organic Carbon (mg kg^{-1})
1.2 m (4 ft)	70:6:24	142
	76:7:17	83
2.7 m (9 ft)	66:5:29	57
4.3 m (14 ft)	67:5:28	42
5.8 m (19 ft)	68:3:29	64
7.3 m (24 ft)	83:3:14	19
8.8 m (29 ft)		

Soil profile showing typical reduction in organic carbon content with depth. Organic carbon increases sorption of contaminants to soil.

Vapor pressure gives an indication of whether the contaminant is likely to volatilize or transfer to the vapor phase from soil or water. This is important from a quality assurance-quality control stand point to recognize the potential loss of sample during sample preparation and analysis. The same property can be used to advantage for monitoring contaminants with high vapor pressures by analyzing the contaminant in the vapor phase without treatment of the soil to remove the contaminants. As with the coefficients for sorption, coefficients have been empirically determined to describe the relative distribution of contaminants between water and air. The Henry's Law coefficient (kPa m^3 mol^{-1}) is a useful tool for comparing the relative volatilities of contaminants from soil in the vadose zone into air spaces of the vadose zone. Henry's Law coefficients also are useful for determining what type of pretreatment, if any, is necessary prior to analysis and if analytical tools to measure gases can be useful for the analysis.

In addition to individual contaminants, often soil contaminants exist in complex mixtures composed of individual chemicals ranging from volatile low molecular compounds to hydrophobic high molecular weight compounds. Petroleum hydrocarbons and wood preserving solutions are examples and are present in many soils from spills, and leaks from underground and above-ground storage tanks and piping. Some individual contaminants and complex mixtures are termed nonaqueous phase liquids (NAPLs). These contaminants such as crude oils are combinations of several chemicals that are nonsoluble or minimally slightly in water. They remain as organic liquids immiscible in water rather than attaching to soil or dissolving in water. They are generally classified as DNAPLs (dense nonaqueous phase liquids) or LNAPLs (light nonaqueous phase liquids). DNAPLs include chlorinated solvents used in the drycleaning industry such as tetrachloroethylene (PCE) and trichloroethylene (TCE). These contaminants are more dense than water and therefore migrate readily through the soil and often reach the ground water. Once they reach the ground water they continue to sink and contaminate large areas of subsurface sediment and ground water. The vertical migration is stopped when a clay confining layer is reach. The DNAPLs pool at the bottom of the confining layer and then can migrate horizontally along the confining layer in an aquifer. Some slow molecular diffusion through the confining layer can occur.

LNAPLs are less dense than water, such as the lighter components of petroleum products including benzene, ethylbenzene, toluene, and meta- and para-xylene (collectively termed BTEX). If they are released on land they also can migrate through soil and reach the ground water. Instead of continuing to sink through the ground water when reaching the water table, these chemicals float on top of the water table and migrate horizontally with the ground water flow. Many of these chemicals are slightly soluble in water. Dissolution of the water soluble components can spread the contamination vertically through soil and sediment. However, this transport is small relative to horizontal transport.

Many NAPLs are regulated in drinking water at $\mu g \ L^{-1}$ concentrations, so even small concentrations in subsurface sediment and ground water are a concern. The NAPL pool acts as a reservoir of contamination through slow dissolution, and continues to contaminate surrounding ground water and sediment for many years. The physical properties of NAPLs make soil monitoring difficult because the sampling locations are critical for determining contaminant plume locations and NAPL concentrations in both sediment and ground water. Also, interactions between individual components of the complex mixtures impact the solubility of the individual compounds and their partitioning between the immiscible liquid and the surrounding ground water and sediment.

Soil Pollution Risk

There is an increasing use of risk-oriented policies to deal with the local effects of soil pollution. The risks that such policies deal with are: Human health risks and can also include eco-toxicological risks. These risks are expressed in terms of negative effects and chances between 0 and 1 that such negative effects will occur. Examples of areas where risk-oriented policies are applied to soil pollution include the United States of America, Canada and countries in the European Union. Historically, these risk oriented policies have followed the abandonment of policies aimed at restoring soils to their original 'clean' state.

Risk-based criteria or standards, developed in the framework of risk oriented policies, are applied to risks estimated with deterministic methodologies, following the steps of hazard characterization, appraisal of exposure and risk characterization, while using exposure-risk relations established beforehand. Risk-based criteria have been applied to decisions about soil remediation in the form of soil clean-up standards, to the use of soils for specific purposes and in the United States also to sediment management. The risk-oriented policies considered here, assume that background exposure to pollutants carries no risk and that a specified level of soil pollution carries a maximum tolerable or maximum acceptable risk for organisms living locally. The latter is the main basis for standard setting.

In part, risk-oriented soil pollution legislation includes policy goals that are qualitative. For instance, the primary UK legislation on contaminated soil defines land as contaminated in need of risk management 'if significant harm is being caused or there is a significant possibility of such harm being caused'. Mostly, however policies have resulted in specific quantitative values for maximum tolerable or acceptable soil pollution. The analysis of such values used in different industrialized countries has shown that there are very large differences, roughly up to a factor. According to Provoost et al., these differences to a large extent originate in different political choices (e.g. including or excluding ecotoxicity) and in different assumptions as to the modeling of exposure to soil pollutants, including site related factors, such as soil type and building constructions.

Risks Related to One Soil Pollutant

In practice, there are several matters which are at variance with the proper establishment of actual risk related to one soil pollutant. These are: The absence of standards for pollutants, neglect of background exposure, and neglect of routes of exposure to soil pollution, neglect of available dose-effect studies and neglect of biological availability.

Absence of Quality Standards

When data regarding soil pollutants are available, they should be compared with quality standards reflecting maximum tolerable risk of exposure. However, such standards are not always in place. For instance, of the volatile organic carbon compounds detected in groundwater samples by the US Geological Service, were unregulated – with no standards in place. Similarly Patterson et al. found a variety of brominated ethenes in Australian groundwater, all lacking standards.

Neglect of Background Exposure

For a proper estimate of soil pollution related risks, exposure to specific soil pollutants should be evaluated in combination with exposure to the same substance that is not related to local soil contamination. Several countries, such as Canada, Germany, Spain and Belgium, do indeed establish soil clean-up standards while considering background dietary and inhalatory exposure but others, e.g. Sweden, Norway and the Netherlands, do not. Neglecting background exposure or specific types of background exposure may have implications for risk estimates.

Neglect of Routes of Exposure to Soil Pollution

In evaluating exposure to soil pollutants, assumptions regarding exposure routes are important. In this respect difference between countries may be noted. Soil clean-up standards for lead of Norway and Sweden differ in part because in Sweden the dominant exposure route is assumed to be by drinking water and in Norway it is thought to be by drinking water and ingestion of soil.

Inhalation of household dust and soil particles is not always taken into account in governmental decision making about risks of soil pollution. For instance, in the Netherlands inhalation of soil particles has been neglected as an exposure route, but in e.g. Spain it is not. Neglect of inhalation would seem at variance with existing studies. Nawrot et al. have studied the effects of cadmium pollution in soil (around former thermal zinc plants) and found a significant increase in lung cancer risk correlated with cadmium exposure. They plausibly explain this in terms of exposure of lung tissue to cadmium present in inhaled soil and household dust particles.

Household dust particles have also been found to be important in the exposure of children to pesticides in agricultural settings.

Neglect of Available Dose-effect Studies

Akesson et al. have analyzed the effects of low environmental cadmium exposure in an epidemiological study of Swedish women in the Lund area, being 64 years of age, excluding women from areas with soils heavily polluted by cadmium. Akesson et al. found associations between the internal dose of cadmium and tubular and glomerular kidney effects, which may represent early signs of adverse effects. Women with diabetes seemed to be at increased risk of experiencing such early signs. In view of these data it seems plausible that at a background exposure that is common in Sweden, old women in the general population may be at risk for adverse cadmium effects and that even a modest increase in cadmium exposure due to polluted soil may lead to added risk. However, when establishing soil clean-up standards in Sweden this background exposure has been neglected. Nawrot et al. have studied the relation between mortality and cadmium body burden in Belgium. They obtained evidence that total mortality and non-cardiovascular mortality may be elevated at cadmium body burdens which can be found among the population not living on soils that are currently considered to be a health risk. Similarly there are now strong indications that the negative effects of lead on the neurophysiologic and sexual development may well be found at the level of background exposure common in Western European and US cities, though soil pollution policy, at least in European counties, assumes that such background exposure is safe.

Ecotoxicological Risks

Maximum acceptable or maximum tolerable ecotoxicological risks are usually derived from a limited number of studies concerning single species under laboratory conditions. Laboratory conditions may be very different from actual conditions in the field, and thus findings in the field are often at variance with laboratory studies. In field studies it has been found that several factors which tend to be neglected in laboratory studies may strongly impact toxic effects of soil pollutants. These include among others: density and adaptability of populations of affected organisms, the presence of other environmental stress factors and the presence or absence of specific landscape elements such as buffer strips.

Biological Availability

Biologically available pollutants determine risk. Biological availability may vary strongly for different types of organisms. Biological availability of a compound in a specific soil is also dependent on physical, chemical and biological and spatial factors. Examples of such factors are pH, the amount and nature of organic and mineral compounds also present and the presence of organisms that can mobilize soil pollutants. In practice, biological availability may be much at variance with total concentrations.

Combination Effects

Limited Accounting of Combination Effects

As to the overall risk of soil pollutants, cumulative effects of the combination of substances present in soils should be considered. However actual standard setting practice has largely focused on criteria relating to one element or compound. In some cases there are criteria for groups of compounds. Such criteria limit the amount (in g/kg soil) of groups of compounds but often do not address the possibility that the risk per unit of weight may be different for different compounds. An exception to this is criteria for the presence of halogenated dioxins and benzofurans and planar biphenyls. The establishment of risk in case of exposure to these compounds uses addition on the basis of equivalent toxicity. This is a major improvement, though it has been pointed out that this approach may still underestimate the risk of neurodevelopment effects.

Importance of Combination Effects

Combination effects may be important in two respects. Firstly, coexisting soil contaminants may impact each others' biological availability. Secondly, exposure to a combination of pollutants may be associated with antagonistic, synergistic and additive interactions of these pollutants, impacting their effect on organisms. Some risks of pollutant mixtures can be predicted on the basis of existing knowledge. For instance there is a fair chance that there will be dose additivity when effects are receptor mediated. Also in case of narcotic effects, joint-mixture ecotoxicological effects may be predicted. If responses are dissimilar, response addition may be used. A methodology to deal with the ecotoxicity of mixtures giving rise to both dose-additive and response-additive effects has been proposed. This two step model evaluates mixture toxicity for the same mode of action with concentration additivity and the toxicity for different modes of action with response additivity. For determining the severity of ecotoxicological effects in case of heavily polluted soils (in which legal maximum tolerable levels for one or more substances are exceeded), a systematic approach to combination effects based on a mixture of concentration addition and response addition has been proposed.

Remedies for Shortcomings

Remedies would seem possible which would allow for a significant improvement in risk estimates. Unregulated substances can get standards. Standards may be regularly updated on the basis of new dose-effect studies. Risk estimates can include both background exposure and all exposure routes for local soil pollution. Estimates of biological availability can be integrated in risk assessments and improved by better testing of bioavailability or by in-vivo monitoring. The deficiencies in taking account of combination effects in ecotoxicity, discussed in section may be addressed by directly testing of ecotoxicity, when the focus is on ecosystem functioning . However it should be noted that small effects on the functioning of ecosystems may have large effects over time. This necessitates large numbers of replicate tests that may well be beyond routine practice.

In determining combination effects on human health, direct testing on humans is an 'unethical option'. However biomarker-based monitoring of some aspects of soil pollution relevant to humans may be an option. For instance Roos et al. have applied a biomarker based test to original and remediated soils that were contaminated by a variety of polycyclic aromatic hydrocarbons (PAH). They tested the expression profile of cytochromes P 450. Xiao et al. have measured genotoxic risk of soil contamination using an in-vitro assay with Salmonella. Though the relation between such biomarker-based data gathered and the in-vivo risks awaits further elucidation, the application of tests based on biomarkers for soil pollution is an interesting option in dealing with combination effects on humans.

Also, estimates of risk may be derived from biomarkers which may be monitored in people exposed to soil pollution. Such biomarkers have emerged from epidemiological studies considering the combined effect of substances. An illustration thereof is the study by Lee et al. which found a graded association of the concentration of blood lead and urinary cadmium concentrations with oxidative stress related markers in the

US population. This suggests that oxidative stress may be useful as a biomarker for combination effects. It has furthermore been proposed to evaluate effects of exposure to nitroarenes by measuring haemoglobin adducts, and of mixtures of volatile organochlorines by measuring glutathione conjugative metabolites. Bioassays based on aryl hydrocarbon (Ah) receptor mediated mechanisms have been proposed which will allow a better alternative to the measurement of polyhalogenated aromatic hydrocarbons. Another option is to estimate risks to human health by taking into account cumulative combination effects in line with established cause-effect relations and research into the effects of actual combinations. It has been shown that risks of compounds with the same targets and the same modes of action may be estimated on the basis of concentration addition, while including toxicity equivalence factors for the compounds involved.

This has been shown to apply to receptor-mediated-and reactive mechanisms of toxicity, provided that no chemical reactions occur between the components of the mixture considered. Currently this approach is applied to halogenated dioxins, benzofurans and planar polybiphenyls, though non-linear interactions are not completely absent in this category of compounds, and neurodevelopment effects may be underestimated, as pointed out before. Extension of this approach is possible to e.g. polycyclic aromatics, including heterocyclic polycyclic aromatics organophosphates that inhibit the enzyme cholinesterase, compounds that bind to estrogen receptors, carcinogens, a variety of petroleum products and compounds that inhibit the MXR efflux pump.

Ecological Risk Assessment

Ecological risk assessment (ERA) is a process of collecting, organizing, and analyzing envi- ronmental data to estimate the risk or probability of undesired effects on organisms, popula- tions, or ecosystems caused by various stressors associated with human activities. All varieties of ERA are associated with uncertainties. The value or usefulness of the different ERA methodologies depends on the uncertainty, predictability, utility, and costs. There are typically two major types of ERA. The first is predictive and is often associated with the authorization and handling of hazardous substances such as pesticides or new and existing chemicals in the European Union. This kind of ERA is ideally done before environmental release. The second type of ERA could be described as an impact assessment rather than a risk assessment, as it is the assessment of changes in populations or ecosystems in sites or areas already polluted. The predictive method is based on more or less generic extrapolations from laboratory or controlled and manipulated semi field studies to real-world situations. The descriptive method is more site specific as it tries to monitor ecosystem changes in historically contaminated soils such as old dumpsites or gas facilities or in field plots after amendment with pesticides or sewage sludge, for example.

Often ERA is performed in phases or tiers, which may include predictive as well as descriptive methods. The successive tiers require, as a rule of thumb, more time, effort, and money. The paradigm or schemes for ERA may vary considerable from country to country, but often consist of an initial problem formulation based on a preliminary site characterization, and a screening assessment, a characterization of exposure, a characterization of effects, and a risk character ization followed by risk management. Although exposure assessment is often just as or even more important. In most European countries, ERA of contaminated soils consists of rather simplified approaches including soil screening levels (SSL) (a.k.a. quality objectives, quality criteria, benchmarks, guideline values) and simple bioassays for a first screening of risk. National research or remediation programs have led to the development of a large variety of guideline values.

Although hard to categorize, most fall into two categories: generic or site specific. While the site-specific guidelines require a characterization of pH, organic matter, etc., at the site, generic guideline values are more independent of modifying factors and hence straightforward to legislate. Three major classes of tools for assessing ecological effects may be identified: standardized ecotoxicity experiments with single species exposed under controlled conditions to single chemicals spiked to soil; ex situ bioassays, here defined as simple laboratory assays where single species are exposed to historically contaminated soils collected in the field; and finally monitoring, analyzing, and mapping of population or community structures in the field. Furthermore, mesocosm, lysometer, or terrestrial model ecosystems (TME) may be useful; these may be considered as large (multispecies) bioassays or ecotoxicity tests. TMEs have the advantage that they operate with the (relatively) undisturbed intrinsic soil populations that make up a small food web. TME hence allow the assessment of effects of toxicants that are mediated through changes in food supply or competition and predation.

One of the keystones in deriving environmental quality criteria is the use of standardized terrestrial test procedures. The emphasis of these prognostic tests is on reproducibility, standardization, international acceptance, and site independence. Although increasing in numbers, relatively few terrestrial tests are still approved by the International Standardisation Organisation (ISO) or Organization for Economic Cooperation & Development (OECD). However, other tests have shown promising results and are likely to be prepared for stand- ardization in the future.

However, the major problem in using simple laboratory tests to extrapolate to contaminated land may not be the limitations of test species and the natural variation in species sensitivity. The problems associated with extrapolating from one or a few species, exposed under controlled and typically optimal conditions, to the complex interaction of species and chemi cals found in most contaminated ecosystems should also cause concern. Although singlespecies laboratory tests with spiked materials have their obvious benefits, e.g., they measure direct toxicity of chemicals and interpretation is therefore simple, supplementary tools are often needed. Bioassays, as defined in this context are one of the more frequently used highertier alternatives. Basically the same test species may be used in bioassays for assessing the risk of a specific contaminated soil as in standard laboratory tests. However, bioassays have the advantage, compared to the use of spiked soil samples, that the exact toxicity of a specific soil may be accessed directly: this includes the combined and site-specific toxicological effect of the mixture of contaminants and their metabolites. Furthermore, the in situ bioavailability of that specific soil is (at least almost) maintained in the laboratory during the exposure period. Several studies have shown a reduction in bioavailability and/or toxicity of soils with an old history of contamination.

Bioassays are therefore often considered a more realistic tool than generic soil screening levels based on spiked laboratory soils. However, a number of uncertainties or problems may be associated with the use of bioassays and the interpretation of their results. First, the test species are still exposed to the contaminants in a relatively short period compared to the permanent exposure condition found at contaminated sites. Furthermore, they are exposed under more or less optimal conditions, in that stressors such as predation inter- and interspecies competition, drought, frost, and food depletion are eliminated during exposure. Finally, typically only a few species are tested individually.

To compensate for some of the limitations just described, contaminated soil may be assessed using multispecies mesocosms, lysometers, or TME. In these, species interactions may be evaluated by manually introducing several species to the systems or monitoring the intrinsic populations of the soil. Natural climatic conditions may be included if the test system is kept outdoors. However, if we want to get a more realistic and large-scale picture of the impact caused by, for example, pesticide use or sewage sludge application, or to assess the environmental health at waste sites, industrial areas, or gas works, it is often necessary to conduct some kind of field observations. Several case studies exist in which field studies have successfully elucidated the ecological risk of specific activities or the ecological impact at specific sites. The small single-species bioassay, large multispecies TME, and field surveys have some drawbacks in common. First of all, it may be difficult to actually link the observed effect to a specific toxic component in the soil. Which of the many substances is actually causing the majority of the observed effects, or is it perhaps a combination of effects? For a hazard classification of soils or a ranking of soils this may not be so important.

However, to evaluate potential risk-reduction measures or risk management procedures it may be important to identify the most problematic substances. A comparison of soil screening values with measured concentrations for each chemical present at a site may be helpful to identify the most likely group of substances causing the observed effect. Other possible tools may include a toxicity identification evaluation (TIE) approach. The TIE approach is a relatively new method, which aims to identify groups of toxicants in soils with mixed pollution. Potentially toxic components present in the soil are fractionated and determined, and the toxicity of each individual fraction is determined by a Lux bacteria-based bioassay or the Microtox bioassay. Although perhaps promising, TIE is a time-consuming and hence costly procedure not yet used routinely. Another crucial issue when analyzing the result of bioassays, TME, and field studies is the presence or absence of a proper reference site or soil. The control soil should in principle resemble the contaminated soil in all relevant parameters, e.g., texture, pH, organic matter, waterholding capacity, and nutrient content, a practical problem that very often is difficult to solve. The lack of adequate control or reference sites may, however, be conquered at least partially by the use of multivariate techniques, which relate the species composition and abundance to gradients of pollutants.

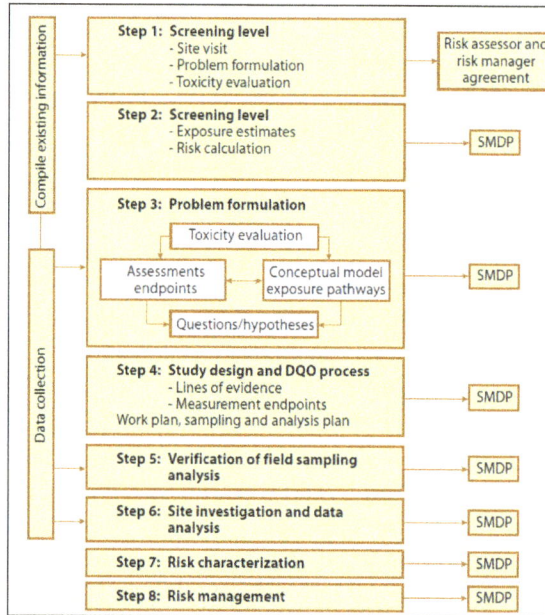

The eight steps in the US-EPA framework for risk assessment of contaminated Superfund sites. DQO = data quality objectives.

However, it is obvious that increased computer power and the presence of new easy-to-use software tools have increased the possibility to move away from more conventional univariate statistics such as analysis of variance (ANOVA) to more powerful multivariate statistics that use all collected data to evaluate effects at a higher level of organization. Statistical methods such as the power analysis may also be very useful in planning and designing large-scale ecotoxicity studies such as mesocosms, TME, or field surveys.

The US-EPA has published an Ecological Risk Assessment Guidance, which should be followed when assessing risks at Superfund sites. As all sites are considered unique this should always be done in a site-specific manner. The ERA process suggested by the US-EPA for Superfund sites follows an eight step process, which can be broken down into four categories, i.e. 1) planning and scoping, 2) problem formulation, 3) stressor response and exposure analysis and 4) risk characterisation. Essential for all steps are a negotiation and agreement of the need for further action between the risk assessor, the risk manager and other stakeholders, the so-called scientific-management decision points (SMDP).

SMDP made at the end of the screening-level assessment will not set an initial cleanup goal. Instead, hazard quotients, derived in this step, are used to help determine potential risk. Thus, requiring a cleanup based solely on those values would not be very likely, although it is technically feasible. There are three possible decisions at the SMDP:

- There is enough information to conclude that ecological risks are very low or non-existent, and therefore there is no need to clean up the site on the basis of ecological risk.

- The information is not adequate to make a decision at this point, and the ecological risk assessment process will proceed.

- The information indicates a potential for adverse ecological effects, and a more thorough study is necessary.

Descision Support System for Ecological Risk Assessment

Ecological Risk Assessment is often a complex process with many variables to take into account. ERA involves many stakeholders and all have to be dealt with in a clear and consistent way. A stepwise or tiered approach is therefore useful to overcome the complexity of an ERA. In order to structure all the information collected, a Decision Support System (DSS) can be used. Each tier will lead to a decision to proceed or to stop. A number of decisions supporting systems or frameworks have already been developed in other countries, e.g. UK, the Netherlands and the USA. The DSS presented here is based on basic principles also common in the methodologies used in the USA and UK. However, in the present DSS measures of bioavailability and the use of the Triad approach may be built into the system more systematically.

Framework for Ecological Risk Assessment

Rutgers et al. developed a basic flowchart for Ecological Risk Assessment, which is used as the backbone of the decision support system (DSS) presented in the figure.

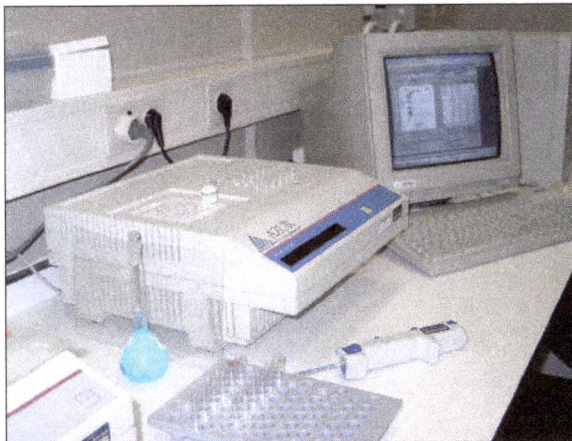

Equipment for measuring luminescence of Vibrio fischer.

The DSS is separated in three different stages, i.e.:

- Stage I. Site characterization and description of land-use.
- Stage II. Determination of ecological aspects.
- Stage III. Site-specific tiered assessment (the Triad):
  - Tier 1. Simple screening.
  - Tier 2. Refined screening.
  - Tier 3. Detailed assessment.
  - Tier 4. Final assessment.

Each of these four tiers is based on a weight of evidence (WoE) approach combining three lines of evidence (Chemistry, (eco)Toxicology and Ecology).

Boundaries of the DSS

It focuses strongly on supporting decisions made when considering risk to the terrestrial environment. Therefore it addresses only indirectly the risk to ground water and associated (connected) fresh water systems. Nevertheless information about e.g. reduced bioavailability may be useful when assessing potential risk for leaching of contaminants to ground water or fresh water. Furthermore, it is important to realise that the management of a contaminated site is more than assessing ecological risk. Issues like for example risk for humans, availability and cost of remediation solutions, development plans for the vicinity or the region are equally important.

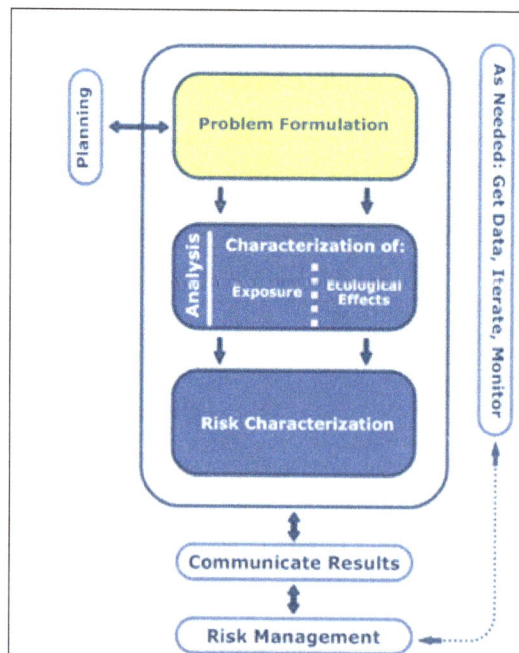

Basic flowchart for ecological risk assessment.

Stage I — Site Characterisation and Land-use Definition

The first step in the DSS is to establish what is often referred to as a Conceptual Site Model. It aims at involving as many stakeholders as possible in order to describe site characteristics and to review all available information from the site, e.g., historical information about land-use, investigation of whether the site may be regulated under specific directives, obvious data gaps and urgency for reaction and data collection. The spatial borders of the site should be defined and the current and the future landuse have to be defined. Consultation between administrators, planners and experts therefore has to take place as early as possible in the process.

a. Initial requirements in the DSS:

An inquiry among all stakeholders should be conducted as one of the first initiatives. The aim should be to collect as much information about soil characteristics as possible.

b. Defining land-use:

One of the first actions to be taken among all stakeholders is to decide which landuse is required for the site, as this will determine the required data collection and testing. Many land-uses may be defined, but generally the four following overall categories of land-use classes are used:

- Industrial area (including infrastructure and pavement).
- Urban/residential area (including recreational and green areas).
- Agricultural area.
- Nature area.

c. When is an ecological risk assessment needed?

Most often a site specific ERA will be initiated only when soil concentrations exceed soil screening levels. However, this may not in itself be a sufficient criterion to go through the entire ERA procedure. Some boundary conditions, based on the present and future type of land-use, the level of contamination and various ecological considerations have to be met in order to rationalize an ERA. The experts and the rest of the stakeholders should answer a number of simple questions in order to conclude whether the required boundary conditions are fulfilled.

Stage II — Determination of Ecological Aspects

At stage II, site-specific ecological features and receptors relating to the land-use defined in Stage I need to be outlined. This includes aspects like key species and life support functions. The potential ecological receptors should be identified in order to determine whether potential source-pathway-receptor linkages can be established. This includes not only ecological receptors directly linked to the site but also those linked indirectly e.g., through leaching of contaminants to connected fresh water systems or (migrating) birds or mammals feeding in the area. Experts from ecotoxicology and ecology should be involved in the selection of ecological aspects.

Stage III — Site Specific Instruments

If after finalising Stage I and Stage II it is still considered that there is a need for a site specific evaluation of ecological risk the process continues to Stage III using the weight of evidence approach described below.

Weight of Evidence Approaches

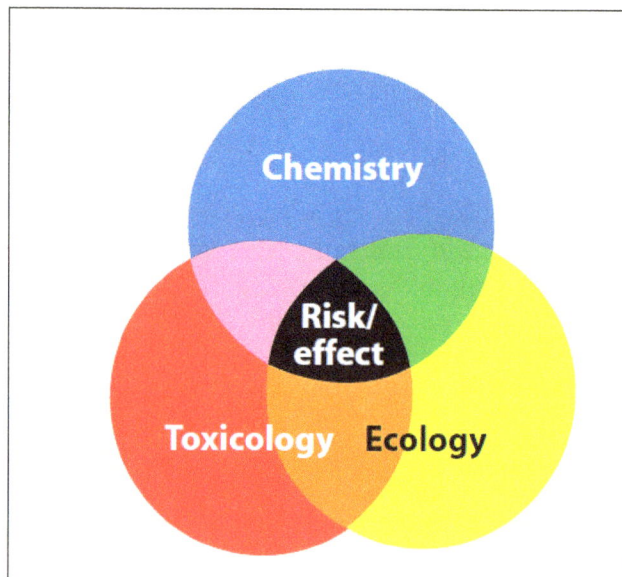

Schematic presentation of the integration of three fields of research according to a Triad.

In order to deal with conceptual uncertainties in a pragmatic way, it has been proposed to use weight of evidence (WoE) approaches for ERA. The rationale is, like in justice, that many independent ways to arrive at one conclusion will provide a stronger evidence for ecological effects, making ERA less uncertain.

In the sediment research area the application of WoE started at an early stage and was called the Sediment Quality Triad. For terrestrial ecosystems WoE approaches and the Triad are still in a developing stage. The Triad approach is based on the simultaneous and integrated deployment of site-specific chemical, toxicological and ecological information in the risk assessment as given in figure. The major assumption is that WoE in three independent disciplines will lead to a more precise answer than an approach, which is solely based on, for example, the concentrations of pollutants at the site. A multidisciplinary approach will help to minimise the number of false positive and false negative conclusions in ERA. It also gives acknowledgement to the fact that ecosystems are too complex to analyse in one-factorial approaches.

- Chemistry: The concentration of contaminants in the environment (totals, bio-available), accumulated in biota, or modelled via food-chains is used for calculation of risks on the basis of toxicity data from the literature.

- Toxicology: Bioassays with species across genera are carried out in order to measure the actual toxicity present in environmental samples from the site.

- Ecology: Field ecological observations at the contaminated site are compared to the reference site. Deviations from the reference site, which can be plausibly attributed to the contamination levels, are funnelled into the Triad.

Using the Triad in Site Specific Assessment of Contaminated Soil

Triad is a powerful weight of evidence approach originally developed in order to evaluate sediment quality. In the terrestrial compartment less experience is available on the practical use of the Triad. Here we describe the use of Traid and give an insight into some of the important decisions risk assessors have to make when conducting the Triad in practise, e.g. how to scale, weight and integrate the outcome of the various investigations.

The Triad approach exists of three lines of evidence (LoE), the so-called Triad "legs", i.e. chemistry, (eco) toxicology and ecology. The Triad approach includes a tiered system in which each consecutive tier is increasingly fine-tuned to the site-specific situation. In the first tier the research is simple, broad and generic. In later tiers more specific and complex tests and analyses may be used. For each of the LoE in the Triad there are a variety of analyses or tests that can be chosen. Some examples are:

- Chemistry: Measurement of total concentrations, bioavailable concentrations, bioaccumulation, etc.

- Toxicology: Bioassays (in field and/or in lab), biomarkers etc.

- Ecology: Field observations of vegetation, soil fauna, micro-organisms, etc. a number of tests or tools that are for suitable for use in each tier are presented for the chemistry, toxicology and ecology LoE.

Decision Charts in Ecological Risk Assessment of Contaminated Sites

Flowcharts

Here we attempt to present a decision support system, which can guide risk assessors in their assessment of site-specific ecological risk. A number of site-specific questions need to be answered before a final decision on performing an ecological risk assessment can be made. Here we introduce a flow chart for ecological risk assessment of contaminated sites. The flowchart is presented as decision trees as shown in Figure together with a more in-depth introduction to the relevant questions that needs to be addressed and answered when performing a site-specific ecological risk assessment.

Decision Making in ERA

The assessment of ecological risk is performed stepwise in tiers. Higher tiers represent gradually more and more complex studies, but also more expensive and laborious studies. The full site-specific risk assessment covers four tiers, i.e.:

- Simple screening: Tier 1.

- Refined screening: Tier 2.

- Detailed assessment: Tier 3.

- Final assessment: Tier 4.

The main principle in going from a simple screening over a more refined screening to a detailed assessment of the contaminated site is to minimize time and effort. The actual performance of the risk assessment and use of the various tiers may be very site-specific.

a. Tier 1 — Simple screening:

After deciding in the two first stages of the ERA that ecological concern needs spe-cial consid- eration, the risk assessment starts typically with a simple evaluation at the screening level. This is done in order to minimize costs until new information indicates the need for further assessment and more sophisticated studies. Therefore, the tools used in the first screening need not only to be reasonably quick and easy, but also rel-atively cheap. The tools for use in Tier I are described in more detail in the toolboxes C T1 and E1. On the basis of the results of instruments used in Tier 1 it is decided to either stop further assessment or continue to a higher tier.

b. Tier 2 — Refined screening:

Tier still considered being at the screening level, aims at refining the measurement of exposure and at the same time to provide further insight into the toxicological and ecological properties of the contaminated soil. Tier 2 deviate from the conservatism normally associated with the use of total concentration in the risk assessment by taking (rough) estimations of bioavailability into consideration in the chemical LoE. A better screening of the toxicological and ecological properties of the soil compensates for the reduced conservatism in the Chemistry LoE of the Triad. The tools for use in Tier 2 are described in more details in the toolboxes C T2 and E2. On the basis of the results in Tier 2 a decision should be made to either stop further assessment or continue to a higher Tier.

c. Tier 3 — Detailed assessment:

The tools in Tier 3 differ from the ones used in Tier 1 and Tier 2 in that they are more laborious, costly and may take longer. On the other hand they are (often) more realistic and/or ecological relevant in order to give a more comprehensive assessment of the ecological risk at the specific site. The stakeholders should beforehand negotiate a minimum set of tests. Is it for example necessary to consider all trophic levels in the toxicological and ecological LoE? Or does the land-use suggest otherwise? Is it necessary (or possible) to estimate the bioavailability of all the substances exceeding their SSL? If not, how are the non-investigated substances dealt with?

The tools described for use in Tier 3 are described in more details in the toolboxes C T3 and E3. Depending on the results from Tier 3 a decision should be made to either stop further assessment or continue with an even more detailed assessment in Tier 4.

d. Tier 4 — Final assessment:

In Tier the aim of the studies is to answer any remaining questions and to decrease existing uncertainties and this may often require more in-depth research. Tools in Tier 4 can be similar to tools of Tier but more focus has to be on site-specific circumstances. For example bioassays should be done with organisms, which normally occur at the site. Furthermore, it may be more relevant to consider ecological effects outside the contaminated area on e.g. predators or herbivores feeding in the area or effects in adjacent fresh water systems. This Tier requires specialised knowledge and experience with ERA, which implies that costs can be high and only a limited number of people may be able to perform the tests. Generally only on a very limited number of site evaluations will include investigations at this level. If the results of Tier 4 still indicate risk there are basically two possible solutions. Accept the risk and leave the contamination or remove (parts of) the contamination.

Screening Tools

Triad based Selection of Methods

For each of the three Lines of Evidence (LoE) in the Triad various methods or tools are available. In order to facilitate the selection of appropriate tools in the right context, the tools have been compiled in subclasses or toolboxes. Each of these is a collection of tools considered to be potentially useful in the designated tiers and LoE of the Triad, i.e. chemistry, toxicology and ecology. Furthermore, the tools are arranged according to their complexity, price and practic ability or in other words depending on whether they are most useful for screening or detailed assessment, i.e:

- Toolbox C1. Chemistry tools for simple screening.

- Toolbox T1. Toxicology tools for simple screening.

- Toolbox E1. Ecology tools for simple screening.

- Toolbox C2. Chemistry tools for refined screening.

- Toolbox T2. Toxicology tools for refined screening.

- Toolbox E2. Ecology tools for refined screening.

- Toolbox C3. Chemistry tools for detailed assessment.

- Toolbox T3. Toxicology tools for detailed assessment.

- Toolbox E3. Ecology tools for detailed assessment.

- Toolbox IV. Various tools for the final (Tier 4) assessments.

a. Toolbox C1 — Chemistry tools for simple screening:

At the very first stage of the ERA process, total concentrations of all relevant chemicals are individually compared to soil screening levels (SSL) in order to evaluate whether there is a need for a site specific assessment of ecological risk. In the current Stage III of the ERA, this first generic evaluation of risk is followed by a more site-specific screening of risk including information from all three lines of evidence in the Triad. In the Chemistry part of the Triad more site-specific information is collected by:

Refining and targeting the comparison of soil concentrations with soil related benchmarks for site-specific purposes.

Incorporation of the accumulative risk of a mixture of contaminants by calculating the toxic pressure (TP) of a mixture and by doing so generating more site-specific insight to the potential ecological impact of a contaminated site. Each of these steps can be done separately or in combination, e.g. the TP can be calculated using existing SSL or using new developed bench- marks based on either NOEC or EC50 values or site-specific benchmarks can be compared to soil concentrations individually. The approach entirely depends on the strategy taken by the stakeholder group and the availability of data.

b. Toolbox T1 — Toxicology tools for simple screening:

The main objective of the selected toxicity tests or bioassay at Tier 1 should be to screen the soil for presence of toxic compounds. This includes toxic degradation products or compounds, which are not routinely included in various national analytical programs for contaminated sites. This Tier is the first screening level of the ERA and the cost in form of manpower and money should hence be relatively low.

c. Toolbox E1 — Ecology tools for simple screening:

Ecological surveys or monitoring studies are generally considered a time consuming effort performed by experts. This is in most cases true, wherefore detailed surveys normally take place in higher tier assessment. However, in order to ensure that also ecological information is collected and used in the Triad already in the screening phase, it is recommended to perform a limited examination of the site. A survey of the area with special focus on visible changes in e.g. plant cover or presence or absence of specific plants, trees or scrubs may indicate ecological damage, which can be associated to contaminants present at the site.

If any aerial pictures areavailable from the area these may give valuable information about the plant cover also historically, which may be helpful in identifying parts of the site where the impact may be highest (hot spots). At this stage the conclusion can in most cases only be indicative. Therefore if the results from the other line of evidence may cause any doubt or the survey indicated potential impact, it is recommended to either continue with a more refined screening in Tier 2 or go directly to the detailed assessment in Tier 3.

d. Toolbox C2 — Chemistry tools for refined screening:

Selective solvent extraction: It may be considered useful to adjust the estimate of exposure by taking bioavailability into consideration and hereby deviating from the conservatism normally associated to the use of total concentration in the risk assessment. The principle in this refinement of the ecological risk assessment is to extract a more ecotoxicologically relevant fraction of the contamination than the total concentration. The latter generally tends to overestimate the risk of historically contaminated soils. In this screening phase no attempt is made to estimate the freely dissolved or readily bioavailable concentration of contaminants. Table 10 explains principal studies that employed chemical extractants to evaluate bioavailability.

Table: Outline of principal studies that employed chemical extractants to evaluate bioavailability.

contaminant	solvent	bioassay	operation	comments
atrazine phenanthrene	methanol/water, n-butanol	Earthworm uptake degradation.	25 ml extractant, 10 g solid. Shaking for 2 h.	Methanol/water best predictor for atrazine, n-butanol for phenanthrene.
DDT,DDE, DDD PAH (mixture)	THF ethanol	Earthworm uptake.	20 ml extractant, 1 g soil, 5 sec mixing.	Good correlation with earthworm accumulation.
anthracene, fluoranthene, pyrene	n-butanol propanol ethyl acetate	plant retention earthworm uptake degradation.	25 ml extractant, 2 g soil, 5 sec mixing.	Reasonable correlation with bioassays.
phenanthrene pyrene chrysene	n-butanol	earthworm uptake degradation	15 ml extractant, 10 g soil, mixing: 5 sec (worm) or 120 sec (degradation)	Applicable for bioavailability prediction.

Instead the fraction of the contaminants is extracted, which can be directly compared to the existing soil screening levels. This is considered to be a relatively simple and quick method to screen for potential risk of contaminants in a more realistic way than using total concentrations. The extracted concentration (mg kg-1) is compared to the

SSL and the result used in the Triad. It is therefore a prerequisite of this comparison that the extractability in the tests (with spiked soils) used for deriving SSL is close to 100% by the methods used. In most short-term tests (< four weeks) it will be reasonable to assume that only little "true" ageing or strong sequestering occurs and hence a majority of the spiked chemicals are still extractable with mild organic solvents. However, for most methods this still has to be fully validated.

Organic solvents most frequently used include methanol/water in different ratios, nbutanol, ethanol, propanpol, ethyl acetate and tetrahydrofuran (THF) given in table. The method establishes preferential partitioning of hydrophobic contaminants to the extractant by increasing their solubility in the aqueous phase whilst removing pollutant compounds from soil surfaces establishing equilibrium conditions. No standard protocol has been adopted for mild chemical extractions in relation to bioavailability testing. Common methodology in literature primarily includes a soil sample to which a volume of chemical extractant is added (generally 1 – 10 g soil,15 – 25 ml extractant). This is followed by a period for mixing, e.g. vigorous mixing for10 – 120 seconds or shaking by orbital shakers for up to 2 hours. The extraction studies have mostly involved PAH and insecticides (including DDT, DDE, DDD and atrazine). Studies that haverelated extractability with results from bioassays have generally focused on uptake and accumulation (% taken up by earthworms or plants) and bacterial degradation (%removed). Therefore, since convincing relationships between the chemical and biological tests were found it may indicate a potential for such extraction methods to predict bioavailability.

a. Toolbox T2 — Toxicology tools for refined screening:

Organisms screening in soil.

In the first simple screening of Tier I focus was on marine bacteria and aquatic/sediment living species. In Tier 2 relatively simple tests with soil dwelling species are used for a more refined screening of the soil samples, i.e. the earthworm survival tests and avoidance tests using soil invertebrates.

The habitat function of soils is often assessed using the reproduction test with Eisenia fetida. The avoidance test with Eisenia fetida is a suitable screening test, which is less cost-intensive in terms of duration and workload than the reproduction test, and at the same time (normally) more sensitive than the acute test with the same species.

b. Toolbox E2 — Ecology tools for refined screening:

In Tier 2 the observations from the survey may be expanded by simple on-site assessment of the overall soil functioning or biological activity of the soils. Recommended tools include baitlamina sticks and simple microbial tests using general endpoints like soil respiration or C/N mineralization rates.

Bait-lamina sticks.

The main principle for tests at this level is to be relatively simple and cheap but at the same time to give valuable information whether or not the soil has lost some of its main services. Bait-lamina sticks for example have been demonstrated useful for describing biological activity of the soils in a general matter.

c. Toolbox C3 — Chemistry tools for detailed assessment:

The objective of the tools found in this toolbox is to assess the bioavailable and freely dissolved fraction of pollutants found in pore water of soils from contaminated sites. The methods should (in principle) be able to mimic the fraction of organic pollutants available for uptake in biota. The collection of methods includes various non-depleting and depleting pore water extrac tions. Very few terrestrial ecotoxicity data are yet expressed as e.g. pore water concentrations. Instead, the outcome of the methodologies in this toolbox is compared with water quality standards.

d. Toolbox T3 — Toxicology tools for detailed assessment:

The objective of the tools found in this toolbox is to evaluate the potential impact of contaminated soils to fauna and plants and hereby the entire ecosystems. Some of the methods use introduced, and not intrinsic, species. The benefit of this is a higher degree of standardisation, as the species used in these bioassays is easy to maintain in laboratory cultures compared to naturally occurring species. The drawback may be that their ecological relevance is less obvious. For example the compost worm Eisenia fetida is used as a surrogate to evaluate risk to soil dwelling earthworms. Two sets of bioassays are presented. One for directly assessing potential risk for soil dwelling species, including micro-organisms, plants and soil inverte- brates, and one for assessing indirectly risk to aquatic species through e.g. leaching of con- taminants. It is often anticipated that soil organisms are exposed to pollutants mainly through uptake from pore water. Therefore it may also be possible to evaluate, or at least to compare or rank, the risk of contaminated soil samples to soil dwelling organisms on the basis of the outcome of the aquatic test using elutriate or pore water. The choice of bioassays depends on a number of variables, e.g:

- The current and future land-use, i.e. targets of protection.

- The size of the contaminated area.

- The potential for ground water or surface water contamination.

- The need of many simple tests or fewer more complicated tests.

Simple plant tests.

e. Toolbox E3 - Ecology tools for detailed assessment:

In this late tier of the Triad, the objective of the activities is community or population response analysis, typically by conducting field surveys. As these studies (most often) are time consuming, costly and dependent on ecologically, taxonomically and statistical expertise they are most frequently done on large-scale sites with a long-term-remediation perspective. In fresh water ecosystem community surveys have been widely used with relative success. The absence of species from places where they would be expected to occur could be a strong identification of unacceptable levels of contaminants. However, this type of studies has only seldom been used for the terrestrial environment. The reasons for this are many. One of the dominants may be the lack of a concentration gradient and obvious "upstream" reference sites at most contaminated areas. No world-wide accepted guideline on how to plan and perform a terrestrial field survey is available and hence no straight-forward and easy-to-follow description can be given. The decision on when, where and how to conduct field surveys depends on a number of issues, e.g. the size of the area, the land-use, the type of contaminants present, time of the year and last but not least the time and money available to perform the study. Nevertheless, a number of general considerations have to be made in the planning phase of a successful field survey. These include (but are not limited to):

- Identify the targets of concern and the species to monitor.

- Elucidate the natural temporal and spatial variation before initiating a field study.

- Use statistical (power) analyses to determine the minimum number of samples or replicates needed to emonstrate the decided difference, e.g. 25% change.

- In order to establish a cause-effect relationship, a number of confounding parameters need to be characterized both at the reference and the test site, e.g. soil type, pH, salinity, hydrology, nutrient and organic matter content and the presence of other contaminants.

As no single description on how to perform ecological surveys for contaminated sites can be given, some general considerations and useful references for this tier of the ecological risk assessment are given below for:

- Assessing impact in the overall biological activity and organic matter breakdown.

- Assessing impact on the microbial community.

- Assessing impact on the plant community.

- Assessing impact on the invertebrate community.

Metabolomics for Soil Contamination Assessment

Soil-based microbes and invertebrates form a vital component of all terrestrial ecosystems. Such communities are important in many steps in nutrient cycles, such as the fixation of nitrogen from the atmosphere and the degradation of decaying matter. However, since many soil-dwelling species cannot be cultured within a laboratory setting, these communities are mostly unstudied, particularly in terms of genetics and biochemistry. Consequently, there has been little insight into how the activities of soil communities as a whole relate to ecosystem functions . We addressed this issue by using principles from the field of metagenomics.

Metabolomics is the generic name assigned to a scientific field that addresses the characterization of low molecular weight organic metabolites released by living organisms in response to environmental stimuli.

The methodological approach of metabolomics relies on a comprehensive analysis of the set of metabolites or "metabolome" produced in response to particular environmental stimuli. Accordingly, the metabolome is the pool of metabolites, small molecules, within a cell, tissue, organ, biological fluid, or entire organism. Exposure of an organism to an external stressor will result in changes at the level of the metabolome, and such changes may constitute a highly sensitive indicator of an external stress. Therefore, metabolomics has potential as a sensitive and rapid technique that can elucidate the relationships between metabolite levels and an external stressor, such as contaminant exposure, nutritional deficit or a disease.

The implementation of metabolomics for the assessment of soil contamination is nevertheless at an early stage.

Methodological Approaches

Metabolites Isolation

Generally, metabolites are extracted from intact organisms (occasionally from selected relevant tissues) that have been exposed to the studied toxicant by moderate chemical extraction. The organisms commonly selected for toxicity testing are earthworms, particularly the genus Eisenia, which are a classic model organism for toxicity assays and have been since long included in official guidelines. Earthworms ingest large amounts of soil and uptake a significant amount of contaminant through the skin. Therefore they are continuously exposed to contaminants. Extractions performed with methanol-chloroform or phosphate buffer solution on pulverized or lyophilised organisms are described to extract the maximum number of metabolites while allowing the performance of reliable analyses.

The isolated extracts usually might not require further sample treatment prior to analysis, which minimizes the introduction of artefacts but also facilitates the development of low cost, rapid methodologies.

Metabolites Determination: Chromatography, Spectroscopy and Spectrometry

The leading analytical techniques in metabolomics for soil contamination assessment are proton nuclear magnetic resonance spectroscopy (^{1}H NMR) and gas chromatography–mass spectrometry (GC-MS, both allowing the identification of compounds at molecular level in the analysed substances. Several authors have also implemented high pressure liquid chromatography (HPLC) and ultra high pressure liquid chromatography (UPLC) coupled with mass spectrometry detector (MS) for the assessment of biological responses to soil contamination with heavy metals. Overall, these analytical techniques allow the determination and identification of the metabolites that foremost represent the metabolic alterations related to the toxic effects of organic or inorganic contaminants in soil.

Metabolomics Data Analysis

The general approach to data analysis in metabolomics can be summarized in three main stages: explorative, supervised and biological interpretation. The explorative phase aims to find groups, clusters and outliers in metabolites and samples studied while the supervised discriminates two or more groups to make predictive models and to find biomarkers. Multivariate methods are currently preferred, although univariate and semi-univariate methods have been commonly used for selecting biomarkers. For instance, the lysosomal system was identified as a particular target for the toxic effects of pollutants in soil organisms. However, it is nonspecific as a marker and only included in a suite of biomarkers among diverse soil invertebrate species can provide the necessary specificity for risk assessment purposes. Finally, the biological interpretation seeks the links between metabolome data and underlying metabolic networks through metabolite set enrichment, pathway analysis and metabolic network inference. Thus, finding metabolite relationships is essential to determine comprehensive and meaningful metabolic changes as biological response to environmental stimuli. Accordingly, such extensive evaluation of the impact of pollutants in the metabolism of target organisms is the approach that can add value to the assessment of soil health and viability of soil organisms undergoing stress from pollution.

Metabolomics Bioinformatics

Information processing by bioinformatics tools and computational biology methods has become essential for solving complex biological problems in genomics, proteomics, and metabolomics. Understanding "omics" data requires both common statistical and computational based methods due to the multi-dimensional and complexity level of the data.

Data-analytical methods for the study of biological systems as developed in the field of computational biology provide a suit of indispensable tools to survey the outcome of metabolomics studies. First, computational biology allows a fast screening of the large biological and chemical data sets generated, and therefore the identification of the most relevant metabolites, i.e. compounds specifically representative of the metabolic changes in the model system following exposure to different concentrations of organic and inorganic toxicants. As a result of the large number of variables (metabolites) studied, metabolomics studies encompass a significant statistical power for the systematic detection of biological responses to environmental changes. Second, the mathematical models developed in computational biology allow the identification of relationships between the external stimuli and the metabolic response. Third, the implementation of computational algorithms to structural biology makes possible to discover the structure-function of new macromolecular compounds, the functional enzymatic conversion and changes in their activity, as well as their molecular interaction and relationship with others compounds in the pathways where they are involved. Moreover, it is possible to detect patterns in such biological responses and establish significant dose-response relationships. Besides, pattern recognition reduces the metabolomics data from hundreds of variables to two or three components that are orthogonal to each other. Overall, this advance of computational biology has been possible due to three significant technological breakthroughs: high-information-content data streams, novel bio-statistical methods, and the computational power to analyse these data.

Data processing and statistical analyses are commonly performed using multivariate (typically a principal component analysis (PCA) and (or) partial least squares (PLS) regression analysis) and univariate (t-test) analyses. These analyses are performed in combination with the quantification and identification of the metabolites. Subsequently, biological interpretation of the data is necessary for understanding the link between the external stimulus and the metabolic response of the organisms.

Principal component analysis is the most widely used multivariate statistical approach in metabolomics, used to explain the overall variability in a data set via a a set of uncorrelated variables called principal components (PCs), which are linear combinations of the original variables. The organization of samples in PCA scores plots is based on the similarities between their metabolic profiles. Thus, PCA allows for dimensional reduction of the data into a low dimensional plane, such as PC1 versus PC2. The scores plot (e.g., PC1 versus PC2) allows for a visual examination of the relationship between the samples based on their metabolic profiles. In a D PCA loadings plot, the contribution (or weight) of each metabolite to the discrimination of the sample classes along one component is represented by the intensity of the metabolite peak. In the D PCA loadings plot discrimination is performed by selecting the points that are scattered further away from the tight cluster of points found near the origin.

Other widely used multivariate statistical tools in metabolomics are PLS regression analysis and PLS discriminant analysis (PLS-DA). Both PLS-regression and PLS-DA are methods for samples classification, with pre-defined variables added to maximize the separation between the sample classes and to construct predictive models. The predefined variables for PLS-regression are measurable quantities such as the contaminant exposure concentration. Validation methods such as the leave-one-out cross validation are used to test the robustness of the models generated by PLS-regression, PLS-DA, OPLS, and OPLS-DA.

Although metabolomics studies mostly use multivariate statistics, univariate statistical analyses can contribute to the information gained from a study. Thus, t-tests can be used to assess the significance of the separation between the controls and stressed organisms in PCA and PLS-DA scores plots. Also, t-tests can be used to determine which metabolites in the ^{1}H NMR spectra of the treatment class increased or decreased significantly relative to the controls.

Biomarkers

The somewhat secondary significance of biological responses for soil contamination assessment was customarily associated to the limitation of biomarkers as measurable responses to contaminants, which classically could only provide an indication of exposure to contaminants in soil. The development of metabolomics, considered an "emerging field" as late as mid-2010 has provided the tools for the determination of multiple biomarkers across different levels of biological organization, and therefore a better assessment of the ecological consequences of contamination. Since the creation of the first metabolomics web database, METLIN, 60,000 metabolites has been incorporated, a rapid development closely related to the evolution of mass spectrometry instrumentation and data analysis tools. Currently, the number of databases and metabolites registered is continuously increasing. For instance, ChemSpider is an aggregated database of organic molecules containing more than 20 million compounds from many different providers. At present the database contains information from such diverse sources as a marine natural products database, ACD-Labs chemical databases, the EPA's DSSTox databases and from a series of chemical vendors. It has extensive search utilities and most compounds have a large number of calculated physico-chemical property values.

One of the goals in bioinformatics is to establish automated and efficient ways to integrate large, biological datasets from multiple sources. This objective is challenging because data sources are heterogeneous in terms of their functions, structures, data access methods and dissemination formats. In addition, the enormous quantity of information produced by "omics" is handled via computers that systematically analyze and store the accumulating sequence, structure and function data. Databases are essential in metabolomics because they provide a rapid and specific tool to identify the compounds isolated from an organism exposed to a particular environmental challenge. Thus, the KNApSAcK package provides tool for analysing datasets of mass spectra as well as for retrieving information on metabolites by entering the name of a metabolite, the name of an organism, molecular weight or molecular formula. A list of metabolites that are associated to a taxonomic class can be obtained by search with the taxonomic name, from which information of individual metabolites can be retrieved. The NIST Chemistry WebBook provides access to chemical and physical property data for chemical species. The data provided in the site are from collections maintained by the NIST Standard Reference Data Program and outside contributors. Data in the NIST Chemistry WebBook can be found by direct searches for chemical species or indirect searches based on related data. Specific databases are also being developed, such as LIPID MAPS, currently the largest database of lipid molecular structures. Otherwise, SetupX combines mass spectrometric and biological metadata, which is a step forward in the organization of information generated by metabolomics analysis.

Metabolomic databases are thus accompanied by accurate description of the biological study design and accompanying metadata reporting on the laboratory workflow from sample preparation to data processing.

Currently, standard analyses focus on the determination of amino acids, mono- and disaccharides, lipids/fatty acids, short chain fatty acids and small phenolics. Accordingly, it is possible to already launch the standardization of metabolomics analysis. For instance, the Northwest Metabolomics Research Center (University of Washington) has established a relevant list of target compounds to evaluate biological responses to changes in the environment. The list of compounds is summarized in Table.

Table: Summary of metabolites and metabolic pathways representative of biological responses to environmental stimuli.

Metabolic Pathways	Number of Metabolites
Alanine, aspartate and glutamate metabolism	15
Arginine and proline metabolism	23
Butanoate metabolism	18
Citrate cycle (TCA cycle)	11
Cysteine and methionine metabolism	14
Fatty acid metabolism	3
Glutathione metabolism	14
Glycine, serine and threonine metabolism	21
Glycolysis / Gluconeogenesis	16
Histidine metabolism	13
Lysine biosynthesis	7
Lysine degradation	6
Nitrogen metabolism	9
Oxidative phosphorylation	6
Pentose phosphate pathway	10
Phenylalanine metabolism	10
Phenylalanine, tyrosine and tryptophan biosynthesis	8
Purine metabolism	30
Pyrimidine metabolism	30

Metabolic Pathways	Number of Metabolites
Pyruvate metabolism	10
Synthesis and degradation of ketone bodies	4
Tryptophan metabolism	15
Tyrosine metabolism	18
Valine, leucine and isoleucine biosynthesis	11
Valine, leucine and isoleucine degradation	5

According to the research results summarized in Table, the implementation of metabolomics in the assessment of soil contamination indicates that contaminants in soil affect several of the major metabolic pathways in living organisms, including glycolysis, trycarboxylic acids cycle and amino acids metabolism. Moreover, data analysis indicates an overall reduction in the production of the associated metabolites. For instance, the interference in amino acids specialized pathways results in a decreased synthesis of purine and pyrimidine nucleotides. These nucleotides are essential for the production of the energy (ATP molecules) that drive most of the enzymatic reactions in living organisms, but also protein synthesis is consequently hampered, which explain the negative effect in processes such as antioxidant activity.

Another emerging group of biomarkers, as highlighted in several studies, are lipids indicate that lipophilic extracts can be used in field based metabolomics experiments to investigate different treatment effects on earthworms. Lipid metabolism is highly sensitive to environmental contaminants, with increasing production of lipoprotein vesicle and lipid peroxidation rate during early stages of the biological response to the presence of a toxicant. Relatedly, earthworm esterases has been proposed as biomarkers for pesticide contamination in soil. Esterases are directly involved in the natural tolerance of earthworms to pesticides, and can therefore be used as specific biomarkers, but furthermore, their characterization by metabolomics approach might help to select the appropriate earthworm species for regulatory toxicity testing. Overall, the increasing specificity of the research performed in ecotoxigenomics will allow a realistic and meaningful incorporation of biological responses in ecological risk assessment.

Oxidative Stress in Contaminated Soil

The induction of the oxidative stress response by the presence of toxic compounds in the environment is a primary mechanisms of defence, although prolonged exposure to contaminants is likely to overwhelm this short-term defence.

Metabolites such as proline possibly detoxify the ROS under stress in vivo. Exposure of plants to both redox active, for example, Cu and Hg, and other metals, for example, Cd and Zn, induces the generation of free radicals that leads to oxidative stress. This represents one of the major causes of toxicity particularly due to redox metals. The cells are equipped with an elaborate network of antioxidative enzymes and low molecular weight metabolites which mitigate the oxidative stress. Proline scavenges different free radicals in certain in vitro generation and detection systems.

Proline quenches ROS and reactive nitrogen species (RNS), which relieves the oxidative burden from the glutathione system. Moreover, polyamines also have an antioxidative role by quenching the accumulation of O_2^- probably through inhibition of NADPH oxidase. This may facilitate phytochelatin synthesis and enhance metal tolerance.

Overall, oxidative defence response to toxicity or other environmental stress involves the generation of oxygenated metabolites from exposed organisms and activation/inhibition of the production of antioxidants enzymes and metabolites such as glutathione. The depletion of antioxidants for prolonged exposures might result in the decrease of the response effectiveness and eventual imbalance between generation and elimination of reactive oxygen species. Depletion of glutathione appears to be a major mechanism in short-term heavy metal toxicity. In accordance with this hypothesis, a good correlation between glutathione contents and tolerance index was observed with 10 pea genotypes differing in Cd sensitivity. High GSH concentrations in hyperaccumulator T. Goesingense coincided with high constitutive activity of serine acetyl transferase (SAT); SAT catalyses the acetylation of L-Ser to OAS which in turn provides the carbon skeleton for Cys biosynthesis. Elevated GSH levels in T. Goesingense also coincided with the ability both to hyperaccumulate Ni and to resist its damaging oxidation effects.

The significance of glutathione and the metal-induced phytochelatins (PCs) in heavy metal tolerance has been studied intensely. However, PCs are important for detoxification of only a limited set of metals such as Cd^{2+}, Cu^{2+} and AsO while Zn^{2+} and Ni^{2+} are poor inducers of PCs and exhibit low binding affinity. Most other metals lack significant binding.

Evaluation of metabolites related to oxidative response constitutes a relevant group of target compounds for risk assessment. Although oxidative response to soil contamination has been classically addressed in plants, the study of this response in soil microorganisms is already being introduced in ecotoxicology as a fundamental part of the biological response of soil microorganisms to soil contamination. Accordingly, describe the attenuation of the oxidative response for springtails in laboratory tests, which constitutes and early detection of soil pollution, and standardized test have been developed.

Metabolites Related to Soil Contamination with Organic Compounds

It is possible to infer that soil contamination with organic compounds, namely pesticides o polycyclic aromatic hydrocarbons, abates essential metabolic pathways such as the trycarboxylic acid cycle and the oxidative stress response, while lipid metabolism appears to be enhanced. However, the advance in the application of bioinformatics is providing further progress in terms of identification of specific biomarkers for risk assessment of individual target compounds. Thus, toxicity of endosulfan has been directly related with alterations of the GABA-glutamine cycle, while chlorpyrifos depresses the Cori cycle and reduces the production of phospholipids, as indicated by lower levels of choline specifically relates chlorpyrifos toxicity to increased levels of fumarate, an intermediate of the trycarboxylic acid cycle. Research conducted with the same earthworm (E. fetida) and other families of organic compounds revealed a different metabolic response, confirming the capability of metabolomics to discriminate the metabolic pathways involved in the response to a particular toxic compound. Moreover, the results strongly suggest that sets of biomarkers might be soon sufficiently reliable as for their implantation in in toxicity standardized test.

The relevance of these and future studies on the development of risk assessment strategies is aggravated by the inherent risk of soil contamination for human health. Soil contaminants may be responsible for health effects costing millions of euros. Health problems range from cancer (arsenic, asbestos, dioxins), to neurological damage and lower IQ (lead, arsenic), kidney disease (lead, mercury, cadmium), and skeletal and bone diseases (lead, fluoride, cadmium).

Overall, few studies have been conducted on the toxicity of complex chemical mixtures in soils. The effects of the soil and organisms within it upon organic pollutants are unknown. The data currently available correspond mostly to short-term studies and high level exposure of these chemicals, which is less relevant to the potential low-level, long term health impacts on living organisms near to contaminated soil.

Metabolites Related to Soil Contamination with Heavy Metals

The uptake of excess metal ions is toxic to most organisms, and the biochemical impact of metal ions on the cells varies with the chemistry of the element as their chemical nature. In plants, phytotoxicity of heavy metals in most parts can be attributed to symplastic accumulation of heavy metals, such as the cytosol and chloroplast stroma. Metal-induced changes in development are the result of either a direct and immediate impairment of metabolism or signaling processes that initiate adaptive or toxicity responses that need to be considered as active processes of the organism. Transport processes have been recognized as a central mechanism of metal detoxification and tolerance.

Some metals, for example, Zn and Cu, are essential for normal plant growth and development as they serve as structural and functional components of specific proteins. Other metals, for example, Cd and Pb, have no known function in plants although a Cd requirement for carbonic anhydrase from marine diatoms has been reported.

Upon exposure to metals, organisms often synthesize a set of diverse metabolites that accumulate to concentrations in the millimolar range, particularly specific amino acids, such as proline and histidine, peptides such as glutathione and phytochelatins (PC), and the amines spermine, spermidine, putrescine, nicotianamine, and mugineic acids that can be detected as response to these metals exposure. The advance of toxicogenomics in relation to organic contaminants is significantly ahead of the equivalent research in metal contaminated soil. Nevertheless, research conducted up to date has yielded a number of biomarkers representative of the biological response of soil microorganisms to metals toxicity. Thus, soil contamination with Pb has been related with an enhancement of lipid metabolism and more directly with reduction of tyrosine levels. Otherwise, Cd toxicity promotes the secretion of phytochelatins in C. elegans, likely at the expenses of the sulphur metabolism, as suggested by the reduction in cystathionine, while the response of tomato plants to Cd involves several biochemical pathways. These examples illustrate the genuine specificity of biological reactions to different metals but also the variation in representative biomarkers among different organisms. Accordingly, exposure of C. elegans to Ni yields a different metabolome than Cd since different biochemical pathways are affected.

In plants, data currently available demonstrate the significance of nitrogen-containing metabolites beyond phytochelatins and glutathione in plant response and acclimation to heavy metals. The various metal ions have specific chemical properties and induce distinct responses of adaptation and damage development. Thus, accumulating N-metabolites display a variety of functions, i.e. metal ion chelation, antioxidant defence, protection of macromolecules, and possibly signalling.

Proline is an extensively studied molecule in the context of plant responses to abiotic stresses. Up-regulation of proline is often encountered in plants under heavy metal stress, comparable to what occur under other abiotic stresses. When compared at equal toxic strength, proline accumulation decreased in the order Cd > Zn > Cu. In addition, it has been suggested different functions of proline under metal-stress, being involved in osmoregulation, metal chelation, antioxidant, and regulator of specific functions in plant morphogenesis.

Furthermore, Ni-hyperaccumulation has been specifically linked to histidine production, particularly for Saccharomyces cerevisiae. The beneficial role of high histidine levels has been shown in transgenic Arabidopsis thaliana which accumulated about fold higher histidine levels than wild-type plants and showed more than fold increased biomass production in the presence of toxic Ni in the growth medium. Moreover, cell surface-engineered yeast displaying a histidine oligopeptide (hexa-His) has been shown to adsorb 3–8 times more copper ions than the parent strain, being more resistant to Cu than the parent.

Otherwise, polyamine contents are altered in response to the exposure to heavy metals. Weinstein et al. showed an increment in putrescine content in Cd-treated oat seedlings and detached oat leaves with a marginal rise in spermidine and spermine content. They influence a variety of growth and development processes in plants and have been suggested to be a class of plant growth regulators and to act as second messengers. It has been suggested that they could stabilize and protect the membrane systems against the toxic effects of metal ions, particularly the redox active metals.

References

- Soil-contamination-risk-assessment-and-remediation, environmental-risk-assessment-of-soil-contamination: intechopen.com, Retrieved 16 April, 2019

- Metabolomic-analysis-of-soil-communities-can-be-used-for-pollution-assessment-257755595: researchgate.net, Retrieved 14 June, 2019

- Metabolomics-for-soil-contamination-assessment, environmental-risk-assessment-of-soil-contamination: intechopen.com, Retrieved 18 May, 2019

- Belluck, D.A., Benjamin, S.L., Baveye, P., Sampson, J., Johnson, B. 2003. Widespread arsenic contamination of soils in residential areas and public spaces: an emerging regulatory or medical crisis? International Journal of Toxicology 22: 109-128

4

Soil Pollution: Treatment and Control

There are different techniques of soil pollution treatment and remediation. Some of these are excavation and off-site disposal, on-site natural attenuation, bioslurping, bioventing, biosparging, phytoremediation, etc. This chapter closely examines these soil pollution treatment and control techniques to provide an extensive understanding of the subject.

Soil Remediation

Soil remediation, which is sometimes also called soil washing, is a term used for various processes used to decontaminate soil. Healthy soil is better able to grow vegetation, as well as contributing to healthy air and groundwater.

There are a number of different processes for soil remediation, each employing a distinct technique for removing contaminants from soil. However, each has an indicated best use, so care must be taken to select the right method of soil remediation services for each unique application. The best approach is determined with a proper soil sampling.

Thermal Soil Remediation

Thermal soil remediation is a method that removes specific types of contaminants that are best removed by subjecting soil to high temperatures. This process is typically reserved for soil that has been tainted with contaminated water or by hydrocarbon compounds such as oil or other petroleum products. Typically, this takes place in an oven, fed by conveyor belt.

Essentially, the way it works is by baking the soil causing contaminates to evaporate. The extracted materials are captured and cooled for later disposal. The treated soil is then cooled and removed from the remediation machinery via a conveyor system. After the process is finished, the soil is then ready for recycling or further testing.

Encapsulation

This process of soil remediation is somewhat different from other techniques, as most remediation uses a process to filter contaminants from soils; encapsulation ensures they cannot spread any further.

It's akin to a medical quarantine. Instead of treating a disease by giving a patient antibiotics or retrovirals to combat the disease, the patient is isolated to prevent the contagion from spreading further.

The most common technique of encapsulation is to mix the contaminated soil with lime, cement and concrete. This prevents any other soil from coming in contact with the contaminants contained inside. While it is effective, it also precludes using the treated soil for any cultivation of any sort. Therefore, you should not consider encapsulation unless the soil in question is never going to be used in any capacity for growing anything.

Air Sparging

The air sparging method of soil remediation is indicated when soil has been contaminated by toxic gases or vapors. However, it does differ from other methods of remediation in that it has to be applied directly to the soil rather than used on soil extracted for treatment. Air sparging is done by injecting large volumes of pressurized air into contaminated soil or groundwater, removing volatile organic compounds that might otherwise be removed by carbon filtering systems. It's most commonly used for removing hydrocarbon contaminants, but is best applied when the soil cannot be removed first, as it must be done in situ.

Sparging is one of the most common methods of in situ remediation, so this something to consider when looking at a soil remediation method.

Bioremediation

Bioremediation is also an in situ remediation technique, but uses a biological mechanism rather than a mechanical method of filtering for removing contaminants. Contaminated soil is treated in situ by applying engineered aerobic and anaerobic bacterium that feed on the specific type of contaminant that a parcel of soil is contaminated with.

The bacterium goes to work consuming and breaking down the hydrocarbons or other contaminants in the soil. Much like yeast feeding on sugar in a batch of beer, the bacteria die off after the supplies of contaminants are consumed.

However, bioremediation requires specific and stable conditions to be able to work. It works best when a soil temperature of 70 degrees F and only occasional rain. It can work in colder climates if soil is insulated and covered, but will take longer to take effect. It is a very effective method of in situ remediation, but again requires conditions to be amenable to work efficiently.

Remediation Technologies for Contaminated Sites

Contaminated sites are always a public concern for its potential damage to living organisms including human beings, the ecology, the environment, and even property value. If the contamination problem is not adequately identified, recognized, delineated, studied, and practically resolved, it can pose a considerable risk to public health and the environment. It may also adversely affect current or proposed land uses and development of the site. Therefore, it should be carefully considered in the subdivision, development, and redevelopment of land; any change in land use, such as from commercial to residential, or from one form of commercial to another, e.g., from service station to office; where additions and alterations may be proposed to the existing landscape and/or infrastructures; and the transactions of land and properties.

Remediation of soil contamination can be achieved by: (1) in-situ removal of contaminants from the contaminated site for further off-site treatment of the contaminants removed; (2) ex-situ removal of contaminants from the contaminated soil after the soil has been excavated from the contaminated site; (3) in-situ containment of the contaminants with the toxicity of the contaminants remains unchanged but the contaminants are isolated from human contacts for a predetermined period of time; (4) excavation of the contaminated soil and transport it to an engineered containment system for long-term isolation; (5) in-situ transformation of the contaminants so that the mobility and/or the toxicity of the contaminants are significantly reduced so as to reduce the risk of soil contamination to public health and the environment; and (6) any contaminations of these remediation mechanisms. All these mechanisms have their advantages and limitations. Moreover, they are contaminant specific and heavily dependent on the subsurface environmental conditions of the site. Therefore, it is very important to recognize that there is not a single technology or a single combination of technologies that would be applicable to all contaminants under all subsurface environmental conditions.

Subsurface Contamination

Contaminants can exist in different chemical states and different forms in the subsurface. The physical and chemical interactions of contaminants with the existing soil dictate the fate of contaminants in the subsurface. The environmental conditions of the soil depend on the hydrogeology of the contaminated site. The problem is further complicated by the large number of possible geochemical and biogeochemical reactions and the fact that the outcome of these reactions is heavily dependent on the existing environmental conditions, such as pH, Eh (redox potential), salinity, mineralogy of soil solid particles, temperature, pressure, etc. Moreover, these soilcontaminant interactions are dynamic, reversible, and inter-dependent. Nonetheless, these geochemical processes largely determine the feasibility of remediation technologies required if cleanup of the site becomes necessary. Therefore, a thorough understanding of the soil-contaminant interactions under different environmental conditions is essential.

Remediaton Technologies

Excavation and Off-site Disposal

Excavation followed by off-site disposal, often known as dig and haul, is a well proven and readily implementable remediation technology. It is applicable to practically all contaminants. Contaminated soil is physically excavated and transported to permitted off-site treatment and disposal facilities. Pretreatment of the contaminated soil excavated may sometimes be required to satisfy the ultimate land disposal requirements. Prior to excavation and off-site disposal was the most common method for remediating hazardous waste sites in the U.S. Nonetheless, excavation of contaminated soil is the first step of all ex-situ remediation technologies. However, the technology can be expensive when costs of transportation, and off-site treatment and/or disposal are included.

It should be noted that there are hazards implicitly associated with the handling of contaminated soil. The soil may be contaminated with explosive, flammable, or combustible materials, e.g., carbon disulfide, hydrogen sulfide, methane, tetraethyl lead, etc., any sparks generated during excavation may ignite these materials to cause a fire or explosion. Excavation workers may be exposed to volatile organic compounds (VOCs), semivolatile organic compounds (SVOCs), and particulates contaminated with semi-volatile organics, and/or inorganic contaminants. Hazards of inhaling contaminated airborne dusts are particularly evident during warm and dry periods. Workers may also be exposed to the risk of dermal contact with waste materials when handling such materials. Workers may inadvertently ingest contaminants or waste materials that collect on their hands and clothing in the form of dust during excavation, as dust ingestion may occur when workers take water and/or meal breaks, or after they have left the contaminated site if established hygiene procedures, e.g., washing hands, are not strictly followed. Microorganisms in contaminated soil may pose biological hazards at sites containing medical wastes or sewage sludge. Workers may be exposed to inhalation, ingestion and/or dermal contact with pathogens such as Coccidioides sp., Histoplasma sp., Mycobacterium sp., etc.

Other factors that may limit the applicability and effectiveness of the technology include: (1) generation of fugitive emissions may be a problem during excavation and transportation; (2) remediation cost is heavily dependent on the distance between the contaminated site and the nearest disposal/treatment facility with the required permit(s), therefore, the logistics of excavation and transportation of contaminated soil is extremely important in the economic consideration of the technology; (3) depth and composition of the materials requiring excavation must be considered, the remediation technology is typically restricted to shallow soils less than approximately 3 m below ground surface, as backhoes can generally excavate to a maximum depth of approximately 6 m and soil excavation at greater depth requires larger and more powerful earth moving equipment; excavation in the vicinity of infrastructures may require the installation of temporary lateral support systems for safety reasons, which increase the cost and duration of the remediation project substantially; (4) transportation of the contaminated soil through populated areas may affect community acceptability; (5) disposal options for certain waste, such as mixed waste, radioactive or transuranic waste, may be very limited; (6) contaminants can potentially migrate from disposal/treatment facilities through different pathways, such as effluent discharge to surface water, rainfall surface runoff, leaching into groundwater, volatilization to the atmosphere, and dike uptake; and (7) disposal/treatment facilities without proper design and maintenance can develop odor, mosquito, and insect problems.

In-situ Natural Attenuation

In-situ Natural attenuation is a component of all remedial solutions to subsurface contamination. It refers to the natural physical, chemical, and biological processes that reduce the concentration of contaminants in the subsurface. Examples are advection, dispersion, and dilution of contaminants by infiltration; transfer of contaminants in groundwater to the air in soil pores by volatilization; bioremediation of organic contaminants; or reduction of contaminant mobility by sorption onto soil particle surfaces.

In-situ natural attenuation can be an appropriate remediation technology when the contaminants degrade or disperse readily and do not pose a significant risk to public health and/or the environment while they attenuate, in particular when the contamination source has been removed or contained. It is generally not an appropriate technology when (1) the site contains a significant amount of non-aqueous phase liquids (NAPLs); (2) concentrations of contaminants are so high that they pose an unacceptable risk to public health and/or an ecosystem, or become toxic to microorganisms; and (3) the rate of attenuation is unacceptably slow.

In-situ natural attenuation has several advantages and disadvantages in comparison to other remediation technologies. The advantages include: (1) less remediation waste products are generated, therefore the possibility of cross-media transfer of contaminants and human exposure is considerably less than that of most typical ex-situ technologies; (2) less intrusion and few surface operational facilities are required; (3) can be applied to all or part of a site depending on site conditions and remediation goals; (4) can be used in conjunction with, or as a follow up to other remediation technologies; and (5) the overall remediation costs are lower than those of active remediation. However, the potential disadvantages include: (1) remediation time may be longer than that required by a more active remedial solution; (2) characterization may be more complex and costly; (3) degradation of parent compounds may generate more toxic degradation products; (4) long-term performance monitoring is often required; (5) institutional controls may be required for risk management; (6) contaminant migration and/or cross-media transfer may occur if the hydrology and geochemistry of the site change over time; and (7) there may be a negative public perception of the technology, and public outreach and education may be required before the technology can be accepted by stakeholders.

Dispersion, sorption, and volatilization are important processes of in-situ natural attenuation. However, biodegradation is very likely to be the most effective process in most cases. An aquifer is a complex ecosystem that contains a variety of microorganisms competing for food and striving to reproduce. Variables of the system include: temperature, contaminant distribution, pH, soil type, nutrient levels, hydraulic conductivity, geochemistry, and availability of electron acceptors. These variables affect the occurrence and rate of contaminant biodegradation.

In-situ natural attenuation can be enhanced by the introduction of selected chemicals into the subsurface to stimulate or enhance one or more of the natural attenuation mechanisms, in particular biodegradation.

In-situ Containment Systems

Remediation or cleanup of many contaminated sites may be technically impossible, financially unviable, cost-ineffective, impossible to complete within a given time frame, and/or unnecessary when the risk posed to public health and the environment is acceptably low. As a result, attention has been given to the control of subsurface contamination so as to isolate the public from these toxic chemicals. Containment of the subsurface contamination zone using physical barriers may thus be a viable option. Moreover, a contained contamination zone can be used as an in-situ reactor for trials of new in-situ remediation technologies without exposing the public to unnecessary risk associated with unbounded treatment zones.

As shown in figure, in-situ containment systems are composed of: (1) vertical containment barriers of various types (walls); (2) natural or artificially engineered and emplaced bottom containment barriers (floors); and (3) surface containment barriers (caps or covers).

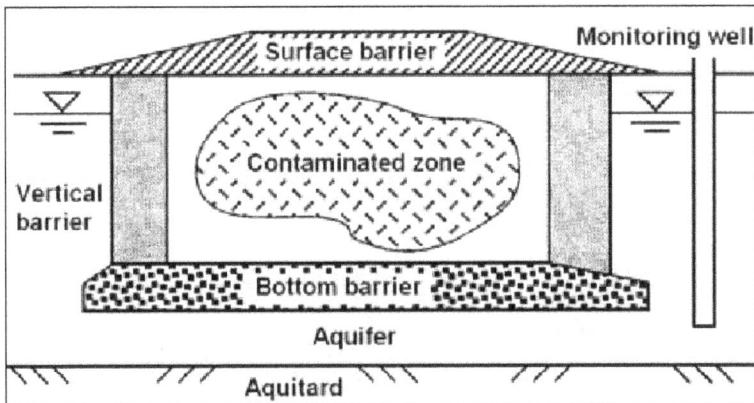

Components of an in-situ containment system.

The functions of walls and floors are to prevent the waste and contaminated groundwater or leachate within the contaminated zone from coming into contact with the and clean groundwater in the surrounding. The functions of caps or covers are to prevent uncontrolled escape of leachate and gases from the contaminated zone; to prevent infiltration of precipitation and runoff into the contaminated zone; to separate the materials in the contaminated zone from humans, animals, and plants; and to serve as the foundation for different types of development and land use atop the contaminated zone.

These containment system components must be compatible to the contaminated materials to be contained, ground deformations, and the effects of cycles of wetting and drying and of freezing and thawing. Design and construction of these containment system components require: (1) definition of objectives; (2) assessment of the existing and future site conditions expected; (3) evaluation of the options; (4) selection of the systems; (5) design of the systems; (6) supervision of construction; (7) post-construction performance monitoring; and (8) cost estimate.

The steps taken in the analysis and design of an insitu containment system to ensure the system will serve the intended functions should include: (1) establishment of the geometry of the containment system; (2) evaluation of the potential movement of adjacent ground; (3) selection of construction materials for the different components of the wall, floor and cap; (4) evaluation of the suitability of potential construction methods; (5) design of joints; (6) cost estimate; (7) development of construction program; and (8) consideration of QA/QC requirements.

Vertical Containment Barriers

Vertical containment barriers are typically constructed as cutoff walls of hydraulic conductivity of $1 \times^9$ m/s or less. When the hydraulic conductivity of the barrier wall is very low or when the hydraulic gradient across the wall is very small, significant contaminant transport by advection is practically prevented. However, it should be noted that migration of dissolved contaminants through the vertical containment barriers can still occur in the absence of advection as shown in figure.

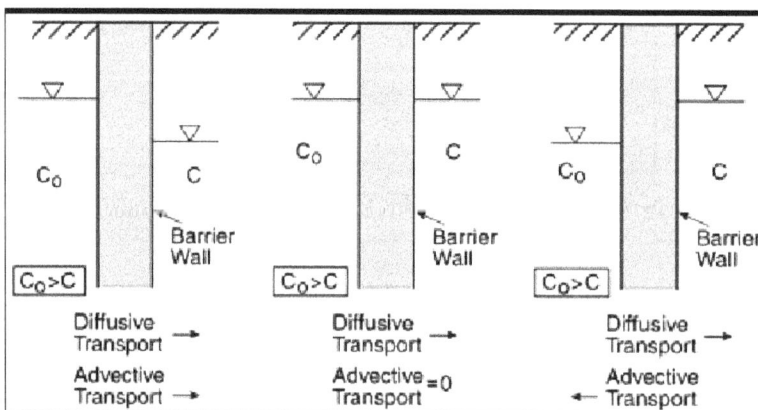

Advective and diffusive contaminant transport across a vertical containment barrier.

When the contaminant concentration in the contaminated zone C_o is higher than that outside the containment system C as shown in figure. molecular diffusion through the pores induced by the concentration gradient may result in significant migration of low molecular weight solutes across soil-bentonite barriers. Therefore, sorbents may have to be added to the barrier to reduce the effective diffusivity of the contaminant in the barrier so as to retard diffusive contaminant transport and to extend the effective design life of a vertical containment barrier.

Soil- and cement-based vertical containment barriers include soil-bentonite slurry trench cutoff walls; cementbentonite slurry trench cutoff walls; plastic concrete cutoff walls; cutoff walls backfilled with mixtures of cement, bentonite, fly ash, ground granulated blast furnace slag, and/or natural clay; cutoff walls constructed by deep soil mixing; and cutoff walls constructed by different grouting techniques. Geomembranes can also be installed as vertical containment barriers.

The selection of the type and materials of vertical containment barriers depends on many factors including: (1) depth and length of vertical containment barriers required; (2) specified maximum allowable hydraulic conductivity; (3) type and extent of contamination; (4) soil types; (5) conditions of bottom containment barriers; (6) depth to groundwater table; (7) local availability of construction materials and equipment; (8) prior local experience; (9) local weather conditions; (10) construction schedule; (11) response of local residents; and (12) costs.

The structural or hydraulic performance of vertical barrier systems can be evaluated by proper quality control and testing specimens excavated from within the barrier constructed. However, less is known about the integrity of vertical barrier systems constructed for environmental remediation purposes. Construction QA/QC, box-outs, pumping tests, injection tests, geophysical evaluation, and post-construction sampling and testing are used for monitoring and performance evaluation.

Bottom Containment Barriers

The bottom containment barriers (floors) underneath a contaminated zone to minimize downward migration of leachate into the environment may be indigenous, i.e., a naturally occurring low hydraulic conductivity geologic formation. The barriers can also be artificially engineered and emplaced, i.e., constructed using admixtures and/or specially engineered construction techniques. They are usually the most uncertain components of an in-situ containment system in terms of their effectiveness as perimeter seal.

The use of native soil and rock strata as the bottom containment barriers depends heavily on the reliable determination of their compositional and physical properties, and the anticipated stability of these properties after prolonged exposure to the contaminants. If the naturally occurring low hydraulic conductivity geologic formation is not a practical solution, it is not always easy to construct the bottom containment barriers below the contaminated zone without disturbing it. However, it can be accomplished by many proven construction techniques such as permeation grouting, non-directional jet grouting to form overlapping cylinders as shown in figure directional jet grouting to form slanted panels as shown in figure. directional drilling and scarifying as shown in figure hydrofracturing and grouting, micro-tunneling techniques, etc. Less proven techniques include ground freezing to control groundwater flow, and the installation of electrokinetic barriers to contaminant transport across compacted clay liners.

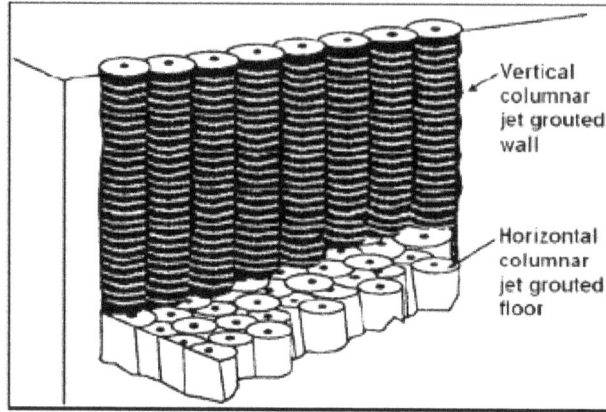

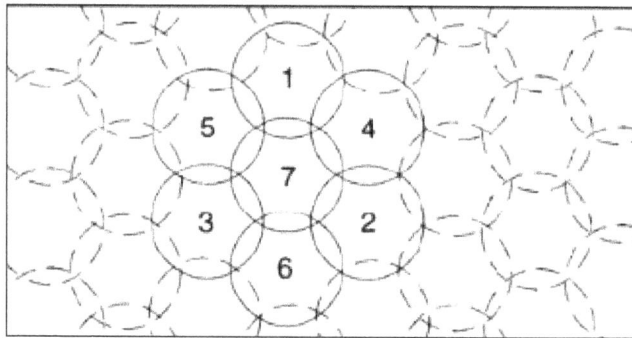

Barrier floor formed by overlapping grout columns.

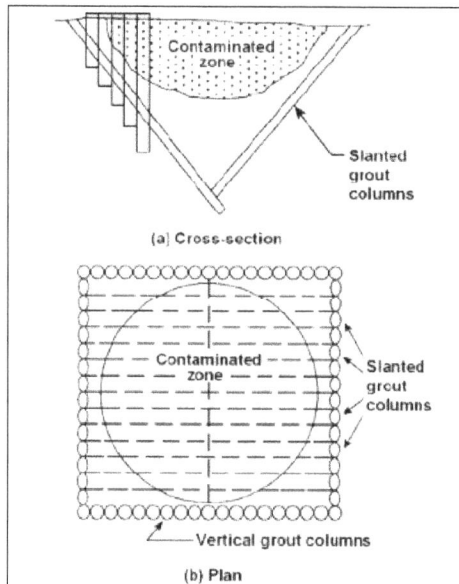

Barrier floor formed by slanted grout columns.

The concept of constructing low hydraulic conductivity bottom containment barriers below a contaminated zone is comparatively new relative to the construction of vertical containment barriers. Whatever quality control and assurance procedures and techniques are employed to evaluate the quality of the constructed barriers, detection of any defects in the constructed barriers, provisions for capture and recovery of any leachate leakage through the barriers, and possibility of repairing the leaky barriers, if necessary, are issues that need to be addressed. In fact, specific future studies have been recommended for the continued advancement of bottom containment barrier technologies by Rumer and Ryan.

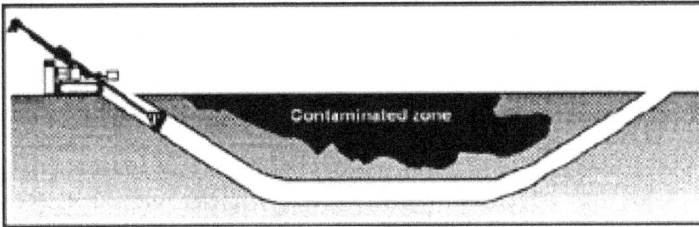

Barrier floor constructed by directional drilling and scarifying technique.

Surface Containment Barriers

The surface barriers (cap) of a containment system may contain six basic components. The functions and typical construction materials of these components are depicted in figure Depending on the characteristics of the contaminants contained in the system, materials available to construct the cap, and site conditions, not all the components shown in figure may be required for all sites. In fact, caps must be tailored to the specific requirements of each particular remediation project. In fact, these components are practically identical to those used in landfill covers with very similar design considerations.

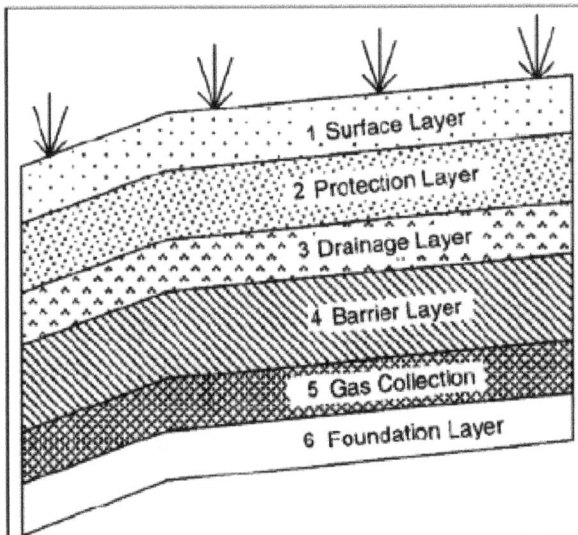

Components of final cap.

Layer no.	Primary functions	Typical materials
1	Separate underlying layer from ground surface; resist erosion; reduce temperature and moisture extremes in underlying layers.	Topsoil; geosynthetic erosion control mat; cobbles; paving material.
2	Store infiltration water before removal; separate waste from humans, animals, and vegetation; protect underlying layers from wetting and drying, and freezing and thawing.	Soil; recycled or reused waste material; cobbles.
3	Reduce water head on barrier layer; reduce uplift water pressure on overlying layers; reduce period overlying layers are saturated after rain.	Sand or gravel; geonet or geocomposite.
4	Impede water percolation through cap; restrict outward movement of gases from waste.	Compacted clay; geomembrane; geosynthetic clay liner; waste material; asphalt.
5	Collect and remove gases.	Sand or gravel; soil; geonet or geotextile; recycled or reused waste material.
6	Foundation for the cap, especially during construction.	Sand or gravel; soil; recycled or reused waste.

The factors affecting the performance of the surface barriers include: (1) types of layers included; (2) materials and thicknesses of individual layers; (3) annual precipitation; (4) surface slope angle of the cap; (5) compressibility of the contained waste; (6) time-dependent deformation of the contained waste; (7) gas generation of the contained waste; and (8) presence of burrowing animals. The design of surface containment barriers thus should consider: (1) design life of the cap; (2) expected routine and major maintenance of the cap; (3) nature of the contained waste; (4) site conditions; (5) materials available; (6) temperature extremes, including possibility of freeze-thaw conditions; (7) precipitation extremes that: (a) affect the design slope angle of the cap surface to promote runoff and control drainage of surface water; and (b) may lead to cyclic wetting and drying that can cause shrinkage cracks in some soils; (8) penetration of plant roots, burrowing animals, worms, and insects into the soil; (9) inadvertent human excavations into the cap; (10) subsidence of the waste, leading to change in drainage contours and formation of cracks in some soils; (11) down-slope slippage or creep, resulting in landslides or cracking; (12) vehicular movements on the cap that can damage the cap; (13) deformation caused by earthquakes; and (14) wind or water erosion of materials.

It should be noted that cap failures are not uncommon and most have occurred during or shortly after construction. Primary causes of failures are: excessive erosion; buildup of excessive pore water pressure in the cap layers; lack of a drainage layer; a drainage layer of insufficient flow capacity; and/or incorrect estimation of the shear resistance between the cap layers. Although most failures did not cause rupture of the barriers, the repair was quite costly.

Stabilization and Solidification

Stabilization is the process by which reagents are mixed with the contaminated soil or sludge to minimize the mobility of the contaminants and/or to reduce the toxicity of the waste. Solidification is the process by which sufficient quantities of solidifying reagents are added to solidify the waste material. The stabilization/stabilization process thus refers to either chemically binding or physically trapping the contaminants in soils. Stabilization/stabilization is used in remediation projects to: (1) improve the handling and physical characteristics of wastes; (2) minimize the rate of contaminant migration into the environment; and (3) reduce the toxicity of certain contaminants. Stabilization/stabilization is a more permanent remedial solution than in-situ containment. It is sometimes referred to as immobilization, fixation, or encapsulation.

The technology may require the contaminated soil or sludge be dewatered or large-size particles or debris be removed or crushed, as there are always operational limits on the maximum size of particles that can be handled by the equipment used to mix contaminated soil with reagents. Contaminated soil or sludge may be treated in-situ, in containers, or in a mobile processing unit. For sites requiring relatively deep in-situ stabilization/solidification or more thorough mixing, modified augers with mixing blades as shown in figure can be employed.

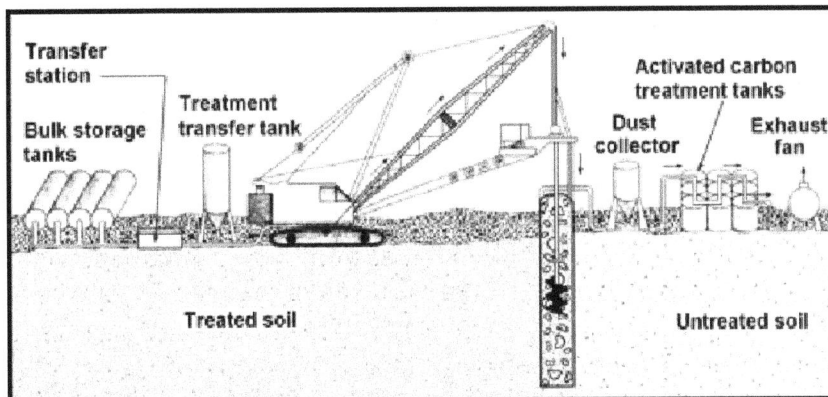

Schematic of an in-situ mixing process with modified augers.

The advantages of the technology include: (1) low cost; (2) applicable to a wide range of contaminants; (3) applicable to different soil types; (4) uses readily available equipment; (5) simple operation; and (6) higher throughput rates than many other remediation technologies. However, the disadvantages of the technology also include: (1) contaminants are not destroyed or removed; (2) the volume of treated soil may be increased significantly by the addition of reagents; (3) emissions of VOCs and particulates may occur during the mixing process, rendering the requirement of extensive emission control; (4) delivery of reagents to the subsurface and achieving uniform mixing in-situ may be difficult; (5) in-situ solidification may limit future uses of the site; and (6) long-term efficiency of the process may be uncertain, in particular when the subsurface environmental conditions change over time.

Reagents used in stabilization/stabilization of waste can be broadly categorized into two types: (1) binder; and (2) sorbent. Binder is a reagent that increases the strength of the product, and sorbent is a reagent that primarily retains contaminants in the stabilized matrix. There is an extensive range of binders and sorbents commercially available, including several proprietary reagents. Some of the non-proprietary reagents include cement; pozzolans such as fly ash, pumice, ground blast furnace slag, and cement kiln dust, lime, soluble silicates, organically modified clays, modified lime, thermosetting organic polymers, thermoplastic materials, etc. Moreover, the waste may also be pretreated by chemical additives to reduce the solubility of contaminants prior to solidification. For example, Cr^{6+} may be pretreated with ferrous sulfate to reduce the chromium to the less soluble and less toxic Cr^{3+}. Arsenic can be immobilized by oxidizing As^{3+} to the less toxic and less mobile As^{4+}, and then treated with ferrous sulfate to form the insoluble $FeAsO_4$. Lead can be pretreated and immobilized by trisodium phosphate Na_3PO_4 to become lead phosphate $Pb_3(PO_4)_2$.

The stabilization/solidification technology employs one or more of these mechanisms: (1) macroencapsulation; (2) microencapsulation; (3) absorption; (4) adsorption; (5) precipitation; and (6) detoxification. However, the stabilization/solidification process is heavily dependent on the type of reagent used.

Cement-based stabilization/solidification is a process that mixes contaminated soil with type I or II Portland cement followed by addition of water, if necessary, for hydration. The hydration of cement forms a crystalline structure consisting of calcium alumino-silicate. The contaminants are bound into the cement matrix and undergo physicochemical changes that reduce their mobility. As a result of the high pH environment generated by cement, the metals are retained within the hardened structure in the form of insoluble hydroxide or carbonate salts. Moreover, acidic waste can be neutralized. The process is applicable to metals, PCBs, oils, and other organic compounds. Extensive dewatering of wet sludges and waste is typically unnecessary as water is required for cement hydration. The disadvantage is the sensitivity of cement to the presence of certain contaminants that may retard or prohibit hydration and the resulting setting and hardening of the material.

Pozzolanic or silicate-based processes involve siliceous and alumino-silicate materials that are not cementitious by themselves. However, they form cementitious substances when react with lime or cement and water. However, pozzolanic reactions are generally slower than cement reactions. The primary immobilization mechanism is the physical entrapment of contaminants in the pozzolan matrix. The treated soil can vary from soft fine-grained material to a hard cohesive material similar to concrete in appearance. The process is applicable to metals, waste acids, and creosotes. Unburned carbon in fly ash may sorb organics from the waste. As a result, a pozzolan such as fly ash may have beneficial effects in the stabilization of both organic and inorganic wastes.

Organic polymerization stabilization/solidification relies on polymer formation to immobilize the contaminants in soil. The process is applicable primarily to special wastes such as radionuclides, but it is also applicable to metal and organic contaminants.

Thermoplastic stabilization/solidification is a microencapsulation process by which the contaminants do not react with the encapsulating material chemically, but is covered with a relatively impermeable layer. For example, the contaminants can be bound into a stabilized/solidified mass by a thermoplastic material such as asphalt (bitumen) or polyethylene. The process is applicable to metals, organics, and radionuclides.

A variety of tests is being used to evaluate the longterm stability of the stabilized/solidified waste. The primary objective of the stabilization/solidification technology is to reduce the rate of contaminant migration into the environment. However, contaminants in the treated waste can migrate into the environment as leachate when precipitation infiltrates the treated waste. The fluid to which the contaminants are leached is called the leachant. After the leachant has become contaminated, it becomes leachate. The overall ability of a treated waste to leach contaminants is denoted as leachability. Available tests used to gauge the leachability of the treated material include: (1) paint filter test; (2) liquids release test; (3) extraction procedure toxicity characteristics (EPTox); (4) toxicity characteristic leaching procedure (TCLP); (5) modified uniform leach procedure; (6) maximum possible concentration test; (7) equilibrium leach test; (8) dynamic leach test; (9) sequential leach test; and (10) multiple extraction procedure. It should be noted that the test method affects the leachability of the specimen determined experimentally, in particular, test variables that affect the contaminant concentrations in the leachate include: (1) leachant-to-waste ratio; (2) surface area of the waste; (3) type of leachant; (4) pH of the leachant; (5) contact time; (6) extent of agitation; (7) number of replacements of fresh leachant; (8) extraction vessel; and (9) temperature. The impact of these test variables should be self-evident. It should also be noted that the selection of the type of chemical analysis and the analytical procedures for the leachate is not a trivial exercise.

Additional physical and engineering property tests are required to evaluate the physical integrity and engineering properties, such as strength, compressibility, and hydraulic conductivity, of the treated material. These tests include: (1) moisture content; (2) bulk and dry unit weight; (3) specific gravity of the solid component; (4) particle size distribution; (5) laboratory cone index; (6) pocket penetrometer; (7) microstructural examination; (8) supernatant formation during curing and rate of setting; (9) unconfined compressive strength; (10) consolidation characteristics; (11) hydraulic conductivity; (12) wet/dry durability; and (13) freeze/thaw durability.

Pump-and-treat

Pump-and-treat remediation is to extract groundwater from contaminated site and process it through a water treatment system. The pump-and-treat technology relies on the advection of water through the contaminated zone, transfer of contaminants to the water, and extraction of the water from the subsurface for further treatment. A typical pump-and-treat system for the remediation of a leaky underground storage tank is depicted in figure.

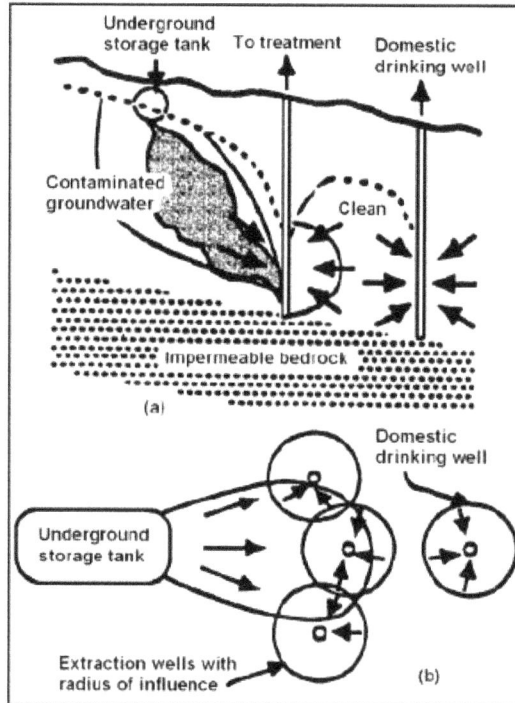

Pump-and-treat system for the remediation of a leaky
underground storage tank: (a) crosssection; (b) plan.

A groundwater pumping system combined with a treatment system, i.e., a pump-and-treat system, is often designed for a specific groundwater contamination problem. The hydrology of the site, the source of contamination, and the characteristics of contaminants must be understood before an efficient and cost-effective pumpand-treat program can be implemented. The system requires the removal of many pore volumes of groundwater for a long period to meet the mandated allowable contaminant levels in the subsurface. As a result, a pump-and-treat cleanup is a relatively slow and expensive process. It usually lasts at least five to ten years, but can last for decades. The time it takes depends on: (1) the type and amount of harmful contaminants in the subsurface; (2) the size and depth of the contaminated aquifer; and (3) the geological conditions of the contaminated site. Cleaning up contaminated water in the subsurface is often very difficult and sometimes not possible. Pump-and-treat is thus the best remediation technology available for such cases. Pump-and-treat can also be used to help keep contaminated groundwater from spreading into nearby drinking water wells while other kinds of cleanup actions are being taken. The U.S. EPA has been using pump-and-treat at over 500 superfund sites.

The contaminant source must first be removed to make the pump-and-treat technology effective. For example, leaking oil drums or tanks must be removed and the contaminated soil in the vicinity must be cleaned up.

An extraction system usually consisting of one or more wells equipped with pumps is built to extract contaminated groundwater from the subsurface. The pumps draw the contaminated groundwater into the wells and up to the ground surface where the water goes into a holding tank and then to a treatment system for treatment. Treatment technologies for the contaminated groundwater extracted can be grouped into three broad categories: physical, chemical, and biological. Physical treatment methods include adsorption, density separation, filtration, reverse osmosis, air and stream stripping, and incineration. Chemical treatment methods include precipitation, oxidation-reduction, ion exchange, and neutralization. Biological treatment methods include activated sludge, aerated surface impoundments, anaerobic digestion, trickling filters, and rotating biological discs.

When a significant amount of free product of a NAPL exists in an aquifer, the system must be designed to maximize the recovery of the free product. Caution must be exercised during the recovery of light non-aqueous phase liquid (LNAPL) when an extraction well is used to control the local hydraulic gradient so as to collect the free product in the cone of depression. If the pumping rate is excessive and the cone of depression becomes too deep, residual LNAPL globules can be trapped below the groundwater table due to capillary forces exerted on LNAPL globules by the aquifer material. The LNAPL globules constitute a persistent source of contamination to groundwater after completion of contaminant removal from the aquifer.

The most difficult aspect of pump-and-treat remediation is to achieve efficient extraction of contaminants that are strongly sorbed on the aquifer matrix.

Soil Flushing

Soil flushing is an in-situ remediation technology that uses an aqueous solution to purge or leach contaminants from the soil into the solution. The flushing solution can be plain water or a carefully designed solution, such as a surfactant or cosolvent, that optimizes desorption of contaminants from soil particle surfaces and solubilization of contaminants in the flushing solution.

A schematic of soil flushing is presented in figure. The flushing solution is pumped into the aquifer via injection wells. The solution then flows down-gradient through the contaminated zone where it desorbs contaminants from soil particle surfaces, solubilizes them in the solution, and flushes them towards the extraction wells where the solution is extracted. The contaminated solution is treated using typical wastewater treatment methods and then recycled by pumping it into the injection wells again. The technology can also be applied to treat soil in the vadose zone as shown in figure. The flushing solution is sprayed on the ground surface, which then infiltrates through the contaminated zone by gravitational force to the groundwater. The leachate is finally removed by the groundwater extraction well.

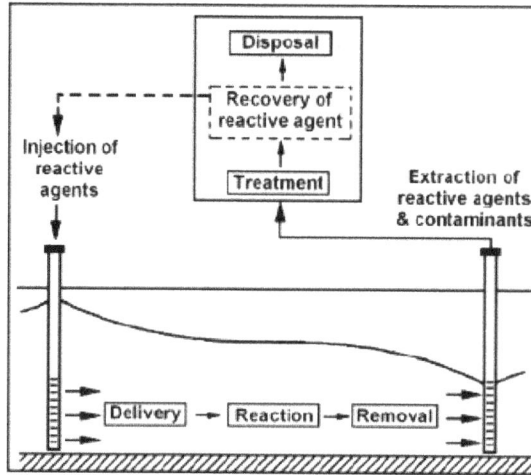

Schematic of soil flushing.

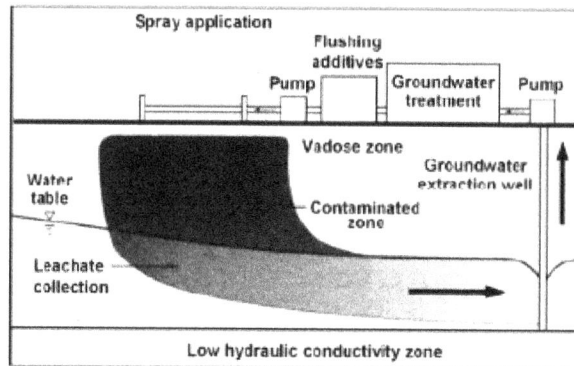

Schematic of soil flushing in vadose zone.

In-situ soil flushing can be applied to many types of organic and inorganic contaminants. The hydraulic conductivity of the contaminated zone is a dominant controlling factor on the applicability of the technology. Regions of hydraulic conductivity higher than $1\times^5$ m/s are considered optimal, and regions of hydraulic conductivity lower than $1\times^7$ m/s are poor candidates for soil flushing. Soils with carbon contents less than 1% by weight are good candidates while soils with carbon contents higher than 10% by weight are generally difficult. Other variables affecting application of the technology include: (1) depth of the contaminated zone; (2) concentration and volume of contamination; (3) distribution coefficients of contaminants between soil particle surfaces and flushing solutions; (4) presence of geologic heterogeneities in the soil horizon; (5) interactions of flushing solutions with contaminated soil; (6) suitability of contaminated site for in-stallation of wells for delivery and recovery of flushing solutions; and (7) design factors such as sizing the delivery and recovery systems to ensure complete recovery of the el-utriate. Flushing of NAPLs may be difficult due to: (1) low water solubilities of NAPLs; (2) high interfacial tension between NAPLs and soil particles; and (3) relatively low hydraulic conductivities of NAPLs because of their relatively high viscosity.

The advantages of the technology include: (1) it is an in-situ technology, therefore it causes less exposure of the cleanup personnel and the environment to contaminants; (2) it is a relatively simple and economical operation to implement; (3) it is applica-ble for a wide variety of contaminants, both organic and inorganic; (4) it is applica-ble for both saturated and unsaturated zone; and (5) it may be used with many other remediation technologies. However, it may have these disadvantages: (1) it may be a slow process when geologic heterogeneities or free products are located within the soil horizon; (2) solubilized contaminants may be transported beyond the influence zone of the extraction well, and unintentional and uncontrolled spreading of contaminants may occur; (3) remediation times may be long; and (4) when the contaminated zone is deep, flushing solution is expensive, and/or remediation time is long, the process may be costly.

The technology can be enhanced using different flushing solutions. Moreover, the tech-nology can be modified by using other innovative remediation technologies simulta-neously. For example, soil fracturing may be used in fine-grained soils to increase in-teractions among flushing solution, soil, and contaminant. In-situ steam injection can stimulate volatilization and solubilization of sorbed contaminants in the subsurface.

Reactive Zone Remediation

The soil flushing technology can be modified to create in-situ reactive zones. The tech-nology focuses on manipulating the in-situ chemistry and microbiology by injecting selected reagents into the subsurface to accelerate natural attenuation of the contam-inant. In-situ chemical oxidation/reduction reactions can then be stimulated to de-toxify the contaminated zone. In-situ reactive zone are applicable for a wide range of contaminants, and they have been applied or are being evaluated on heavy met-als such as chromium, zinc, mercury, copper, arsenic, lead, and cadmium; chlori-nated aliphatic hydrocarbons (CAHs) such as trichloroethene, tetrachloroethene,1,1-trichloroethane, carbon tetrachloride, and daughter products of these compounds; pentachlorophenol; and halogenated organic pesticides such as2-dichloropropane, and2-dibromo-3-chloropropane.

Reactive zones used to remediate subsurface contamination can be created by two dif-ferent methods: (1) by injecting chemical reagents that impact the redox conditions or react with the contaminant in the subsurface; or (2) by injecting electron acceptors and electron donors to enhance microbial growth in the subsurface under aerobic or anaerobic conditions. Depending on the nature of the contamination and the existing subsurface environmental conditions, reactive zones can be oxidizing or reducing.

Oxidizing reactive zones are artificially enhanced subsurface treatment zones where the environment is maintained as strongly oxidizing, i.e., the redox potential is maintained well above 0.0 mV and dissolved oxygen is above 2.0 mg/l. The environment is created by injecting air or oxygen, or chemical oxidants, such as hydrogen peroxide H_2O_2, potassium permanganate $KMnO_4$ or sodium permanganate $NaMnO_4$, ozone, chlorine, or oxygen releasing compounds, into the subsurface. Contaminants are chemically or microbially oxidized. For example, hydrogen peroxide H_2O_2 can be injected with ferrous sulfate to produce hydroxyl radicals to oxidize chlorinated organic compounds into carbon dioxide, water, and chloride ions. Potassium permanganate $KMnO_4$ or sodium permanganate $NaMnO_4$, possibly in combination with H_2O_2, may be used to remediate VOCs, SVOCs, and polychlorinated biphenyls (PCBs). These oxidants have an additional benefit of enhancing a down-gradient naturally aerobic environment to promote bacterial growth so as to accelerate biological oxidation of readily biodegradable compounds such as petroleum hydrocarbons, benzene, toluene, ethylbenzene, xylene (BTEX), and vinyl chloride. However, it should be noted that some geological materials, e.g., pyrite, may release a large amounts of iron and acid when exposed to oxidants. These potential negative impacts must be considered before the application of oxidizing reactive zones.

Reducing reactive zones are artificially enhanced subsurface treatment zones where the environment is maintained as strongly reducing, i.e., the redox potential is maintained well below 0.0 mV and dissolved oxygen is above 1.0 mg/l. Cr^{6+} may be reduced to Cr^{3+} using reducing agents such as Fe^{2+}, Fe^0, calcium polysulfide, or sodium dithionite. Dithionite is a sulfur-containing oxyanion that breaks down rapidly in aqueous solution to form two sulfoxyl radicals. These radicals react rapidly to reduce naturally occurring ferric iron to ferrous iron,

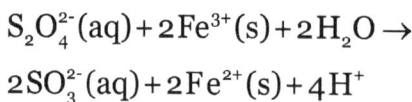

$$S_2O_4^{2-}(aq) + 2Fe^{3+}(s) + 2H_2O \rightarrow$$
$$2SO_3^{2-}(aq) + 2Fe^{2+}(s) + 4H^+$$

Aqueous chromate reacts with Fe^{2+} and precipitates as a solid hydroxide $Cr(OH)_3$ as depicted in equation below,

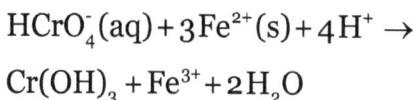

$$HCrO_4^-(aq) + 3Fe^{2+}(s) + 4H^+ \rightarrow$$
$$Cr(OH)_3 + Fe^{3+} + 2H_2O$$

Similarly, hexavalent chromium can also be reduced by ferrous sulfate at neutral pH,

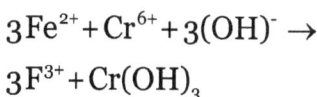

$$3Fe^{2+} + Cr^{6+} + 3(OH)^- \rightarrow$$
$$3F^{3+} + Cr(OH)_3$$

Moreover, cadmium can be precipitated using sodium sulfide and zinc can be precipitated using sodium bicarbonate as depicted in equation below,

$$Cd^{2+} + Na_2S \rightarrow CdS$$
$$Zn^{2+} + NaHCO_3 \rightarrow ZnCO_3$$

In-situ immobilization or fixation can thus be achieved by injection of reagents that transform metal or radionuclide contaminants into an immobile form, such as precipitate.

Aerobic or anaerobic conditions can be enhanced and engineered by a manipulation of the natural environment to create a microbial reactive zone to achieve the remediation goals using in-situ bioremediation and biodegradation of organic contaminants, in particular petroleum hydrocarbons and halogenated aliphatic hydrocarbons.

Design considerations for a reactive zone include: (1) hydrogeology of the contaminated site; (2) groundwater chemistry of the contaminated site; (3) microbiological conditions of the contaminated site; (4) reactive zone layout; (5) baseline definition; (6) regulatory issues; (7) reagents to be used; (8) well design such as type, number, depth, screen zone(s), and layout; (9) reagent feed such as feed rate, solution strength, and frequency of injection; and (10) performance monitoring program.

Soil Washing and Solvent Extraction

Soil washing is an ex-situ technology that uses aqueous solutions to separate organic, inorganic, and radioactive contaminants from excavated soil. It can be used as a pretreatment process to reduce the volume of feedstock for other remediation technologies. The process includes excavation of the contaminated soil, mechanical screening to remove various oversize materials, separation processes to generate coarse- and fine-grained fractions of the contaminated soil, treatment of the individual fractions, i.e., soil washing, and management of the residuals generated. The extracting fluid requires further treatment afterwards to remove and destroy the contaminants. The process also reduces the volume of contaminated soil by washing out fine-grained soil particles where contaminants are sorbed onto and disposed them of as sludge. The technology has been applied successfully to remediate soils contaminated by petroleum hydrocarbons, polycyclic aromatic hydrocarbons (PAHs), polychlorinated biphenyls (PCBs), pentachlorophenol, pesticides, heavy metals, creosotes, and radioactive wastes.

Advantages of the technology include: (1) the technology significantly reduces the volume of contaminated soil as the contaminants are concentrated in a relatively small portion of material, typically 10%; (2) the technology employs a closed system, permitting full control of the environmental conditions, such as pH and temperature, under which the contaminated soil is treated; (3) clean soil after treatment can be backfilled at the same site; (4) potential to remove both organic and inorganic contaminants; and

(5) high throughput rate. However, disadvantages of the technology include: (1) ineffective for soils containing 30 to 50% of silt, clay, or organic matter as contaminants tend to sorb onto these materials; (2) relatively expensive as a result of the additional costs associated with treating wastewater and air emissions; (3) washing fluid may be difficult to formulate for complex mixed waste; (4) small volumes of residual contaminated sludge and wastewater require further treatment or disposal; (5) soil excavation and handling may expose the public and/or cleanup personnel to contaminants; (6) a large space is required to accommodate the soil washing system, system throughput rate, and site logistics.

Solvent extraction technology is similar to soil washing but uses organic solvents to dissolve contaminants and remove them from excavated soil. The technology is based on chemical equilibrium separation techniques being utilized in many industries. It effects the preferential separation of one or more constituents from one phase into a second phase. Typical solvents include liquefied gases, such as propane or butane, and triethylamine. Several additional concerns must be addressed when using a solvent extraction system: (1) solvents must be handled with extreme care as they may be inflammable and of extreme pH; (2) solvent loss during treatment affects emissions to the environment and remediation cost; and (3) amount of residual solvent remains in the treated soil may be of concern due to residual toxicity.

Permeable Reactive Barriers

As shown in figure a permeable reactive barrier (PRB) is an engineered barrier of reactive treatment material placed across the flow path of a contaminant plume in aquifer that removes or degrades contaminants in the groundwater flowing through it. It relies on the natural hydraulic gradient to move groundwater through the barrier. Therefore, there is no continuous input of energy and manpower into the remediation process and there is no mechanical breakdown, thus minimizing long-term operation and maintenance costs of remediation projects. Moreover, technical and regulatory issues relating to the discharge of treated groundwater are avoided or minimized. However, the applicability and effectiveness of the PRB may be limited by: (1) lengthy treatment time relative to other active remediation technologies; (2) potential for losing reactivity of the reactive treatment material, requiring replacement of the material; (3) potential for decrease in hydraulic conductivity of the reactive treatment material due to biological clogging and chemical precipitation; (4) potential of plume bypassing the PRB as a result of seasonal fluctuations in the flow regime; (5) currently limited to shallow depths; and (6) longevity of PRB performance is uncertain.

A PRB can be installed as a continuous reactive barrier or as a funnel-and-gate system. A continuous reactive barrier consists of a permeable reactive cell containing the reactive treatment material. A funnel-and-gate system has an impermeable section, i.e., the funnel that directs the captured groundwater towards the permeable reactive cell, i.e., the gate. The method to emplace PRBs include: (1) conventional excavation; (2)

trenching machines; (3) tremie tube/mandrel; (4) deep soil mixing; (5) high-pressure jetting; and (6) vertical fracturing and reactant sand fracturing.

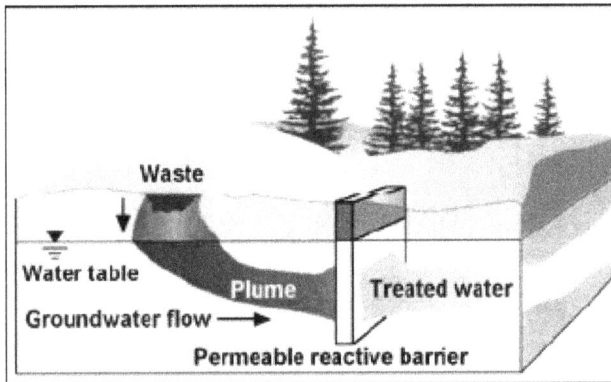

Schematic of permeable reactive barrier.

Permeable reactive barriers use many different treatment mechanisms to remediate contaminants in the groundwater, including: (1) transformation by abiotic and biotic processes; (2) physical removal; (3) pH or Eh modification; (4) metal precipitation; and (5) sorption or ion exchange.

Transformation by Abiotic and Biotic Processes

Crushed limestone has been used in PRBs since 1970s to remediate acid mine drainage and metal-contaminated groundwater. Dissolution of calcite $CaCO_3$, the principal component of limestone, can neutralize acidity, and increase pH, concentrations of alkalinity $\left(CO_3^{2-} + HCO_3^- + OH^-\right)$, and calcium ($Ca^{2+}$) concentration in the mine water. Metal contaminants can precipitate as hydroxides or carbonates, if the pH of the environment is sufficiently high.

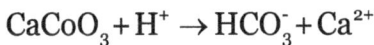

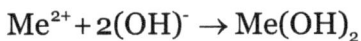

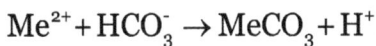

$$CaCoO_3 + H^+ \rightarrow HCO_3^- + Ca^{2+}$$

$$Me^{2+} + 2(OH)^- \rightarrow Me(OH)_2$$

$$Me^{2+} + HCO_3^- \rightarrow MeCO_3 + H^+$$

Where, Me = metal.

The use of zero-valent iron to dehalogenate chlorinated aliphatic organic compounds in groundwater was pioneered by the University of Waterloo in 1992. The degradation process is an abiotic process and the mechanism is attributed to the direct electron transfer on the iron surface. Therefore, the reaction rates are proportional to the surface area of the iron. Zerovalent iron produces a low oxidation potential in groundwater, resulting in precipitation of low-solubility minerals that remove some redox-reactive contaminants such as uranium, chromium, and nitrate.

Corroding zero-valent iron Fe^0 provides electrons by:

$$Fe^0 \rightarrow Fe^{2+} + 2e^-$$

Which can reduce halogenated or nitroaromatic compounds or metals/inorganic species as:

$$RX + 2e^- + H^+ \rightarrow RH + X^-$$

$$M^{n+} + 2e^- \rightarrow M^{(n-2)+}$$

For example, uranium may precipitate as uraninite UO_2 or less crystalline precursor:

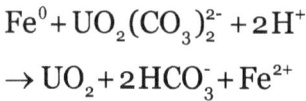

$$Fe^0 + UO_2(CO_3)_2^{2-} + 2H^+$$
$$\rightarrow UO_2 + 2HCO_3^- + Fe^{2+}$$

and Cr^{6+} can be reduced to Cr^{3+}, and Cr^{3+} forms sparingly soluble hydroxide minerals containing a solid solution of Cr^{3+} and Fe^{3+},

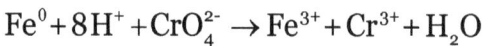

$$Fe^0 + 8H^+ + CrO_4^{2-} \rightarrow Fe^{3+} + Cr^{3+} + H_2O$$

Water can compete for the electrons produced in the corrosion process. In aerobic aqueous systems, Fe^0 reacts with water according to:

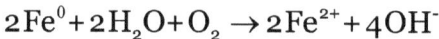

$$2Fe^0 + 2H_2O + O_2 \rightarrow 2Fe^{2+} + 4OH^-$$

Under anaerobic conditions, Fe^0 reacts with water according to:

$$Fe^0 + 2H_2O \rightarrow Fe^{2+} + H_2 + 2OH^-$$

Both of these reactions decrease the redox potential and increase the solution pH within the PRB.

The main criteria for proper functioning of aerobic biological PRBs to treat organic contaminants are availability of surface area and oxygen in the PRB, as bacteria need the surface area to attach and interact with the contaminants in the plume. Oxygen can be supplied in different ways. For example, an air sparging curtain can be installed in the PRB to supply oxygen to enhance degradation of aerobically biodegradable compounds such as BETX. Solid phase oxygen release compounds (ORC) can also be installed in PRBs to enhance degradation of contaminants. Biotic reduction of inorganic contaminants in PRBs has to be supported by supplying electron donor and nutrient materials for microbial growth. Leaf mulch, sawdust, wheat straw, and alfalfa hay have been used as electron donors, and municipal waste or compost has been used as nutrient sources. Dissolved sulfate is an electron acceptor. Oxidation of organic material by sulfate, consumption of acidity, and the coupling to metal reduction are given by:

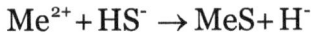

$$2CH_2O + SO_4^{2-} + H^+ \rightarrow 2CO_2 + 2H_2O + HS^-$$

$$Me^{2+} + HS^- \rightarrow MeS + H^-$$

Where, CH_2O = solid organic. The reducing environment generated in the PRB results in the precipitation of metals and other redox-reactive inorganic contaminants. PRBs that employ biotic reduction have successfully remediated metal, sulfate, nitrate, and acid contamination. However, it should be noted that groundwater temperature affects the population of microorganisms and the rate of sulfate reduction significantly.

pH or Eh Modification

Modification of the pH in the contaminant plume is primarily used to precipitate dissolved metals in a plume. A PRB was installed at the Tonolli Superfund Site in Pennsylvania, U.S.A. to remove lead, cadmium, arsenic, zinc, and copper by conveying the groundwater through a limestone bed.

The technique is often used in the first barrier of a multiple barrier system to generate the necessary environment for the reactions in the subsequent barriers. An example of the application is to adjust the pH to within a range in a pre-barrier so that bacteria may grow or bacterial reaction can occur in the biological reaction barrier.

Modification of the Eh of a system involves the exchange of electrons between chemical species that effects a change in the valence state of the participating species. The redox reactions must occur readily under the natural temperature and other environmental conditions of the PRB. Blowes and Ptacek successfully used fine-grained zero-valent iron to generate a highly reducing environment to change the state of dissolved hexavalent chromium to the less soluble trivalent chromium and the metal was then precipitated as chromium hydroxide.

Sorption or Ion Exchange

Sorption is a process by which a chemical in the dissolved phase is sorbed onto solid particles. Most sorption reactions are reversible, pH-dependent, chemical specific, sorbent specific, concentration-dependent, and occur at relatively rapid rates. Surface complexation models are often used to describe the reaction chemistry of various sorption processes. Common sorbents include activated carbon, amorphous ferric oxyhydroxide, zeolites, and ion exchange resins. Most metal contaminants dissolved in groundwater are in the divalent or trivalent state and thus are amenable to ion exchange.

Activated carbon has been widely used to remove organics from groundwater. Its large internal surface area sorbs organics by surface tension. A PRB was installed at the Marzone Superfund site in Georgia, U.S.A. using 815 kg of activated carbon to remediate groundwater contaminated with pesticides including benzene hexachloride (BHC), β-BHC, dichlorodiphenyldichloroethane (DDD), dichlorodiphenyltrichloroethane (DDT), lindane, and methyl parathion. The contaminant concentrations in the treated groundwater are below detection levels. However, it should be noted that the sorption process is reversible. Therefore, the used activated carbon must be removed from the subsurface and managed properly after it has been consumed or the plume is remediated. Otherwise, the sorbed organics can be released to the clean groundwater afterwards. The activated carbon in a PRB thus has to be retrievable. Moreover, the presence of other dissolved ions may compete with organics for sorption sites within activated carbon, reducing the sorption efficiency of activated carbon for organics. Bacterial growth can also foul activated carbon, reducing its effectiveness to remove organics.

Amorphous ferric oxyhydroxide has a high affinity for uranium and metal contaminants. However, the affinity is dependent on the concentrations of carbonate and hydrogen ions in the solution. The predominant uranium removal mechanism in the Oak Ridge and Durango projects may be sorption onto ferric oxides and oxyhydroxides that form from oxidation of zero-valent iron.

Zeolites are hydrated alumino-silicates with a large internal surface area and high sorption capacity. They have been used in PRBs to treat inorganic contaminants by sorption and cation exchange, i.e., a contaminant molecule replaces another molecule at the zeolite particle surface.

Bioremediation and Biodegradation

The most important principle of bioremediation is the use of microorganisms (mainly bacteria) to decompose hazardous contaminants, transform them to less harmful forms, and/or immobilize them under suitable environmental conditions. As bioremediation takes advantage of natural processes, it is thus a very safe technology. The microorganisms existing in soil pose no threat to human at the site or in the community. No dangerous chemicals are used in the process. The nutrients used to make microorganisms grow are fertilizers commonly used on lawns and gardens. When the environmental conditions are favorable, contaminated soil and/or groundwater are remediated in-situ. Therefore, cleanup workers do not need to make direct contact with the contaminated soil and/or groundwater. Moreover, there are practically no harmful byproducts generated by the cleanup process. The process does not need to mobilize as much equipment or labor as most remediation technologies. As a result, the operating cost is relatively low. Bioremediation has been used successfully to cleanup many contaminated sites and is being used at 50 Superfund sites across the U.S.

Although microorganisms live virtually everywhere, their ability to decompose man-made contaminants in the subsurface depends on three factors: (1) types of microorganisms; (2) types of contaminants; and (3) geological and chemical conditions at the contaminated site.

Bioremediation is currently used commercially to cleanup mostly hydrocarbons found in gasoline. However, microorganisms have the capacity to biodegrade almost all organic contaminants and many inorganic contaminants. When microorganisms have access to a variety of chemicals to help them generate energy and nutrients to build more cells, they can do their work to "bioremediate" harmful contaminants. Organic contaminants serve microorganisms in two ways as shown in figure: (1) they are a source of carbon for the building of new cells; and (2) they provide electrons for microorganisms to obtain energy, as microorganisms gain energy by catalyzing energy-producing chemical reactions that involve breaking chemical bonds and transferring electrons away from contaminants.

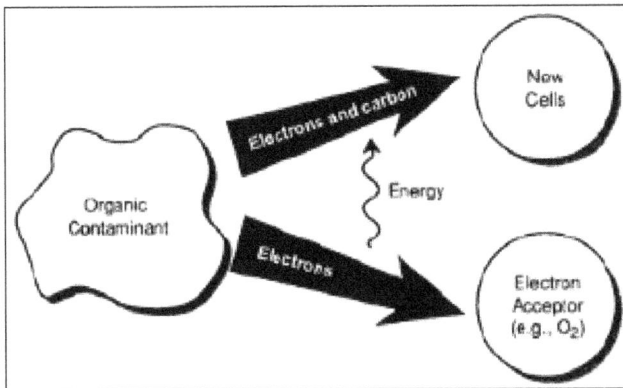

How organic contaminants serve microorganisms.

Under normal environmental conditions, many microorganisms use molecular oxygen O_2 as the electron acceptor, and the process of decomposing organic contaminants using O_2 is called aerobic respiration. Microorganisms use O_2 to oxidize part of the carbon in contaminants to carbon dioxide CO_2 to generate energy, and the remaining carbon to produce new cell mass. In the process, O_2 is reduced to water H_2O. Therefore, the by-products of aerobic respiration are carbon dioxide, water, and an increased population of microorganisms. The biochemical reaction is depicted by:

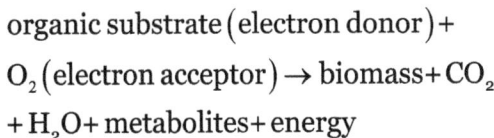

organic substrate (electron donor) +

O_2 (electron acceptor) $\rightarrow$ biomass + CO_2

+ H_2O + metabolites + energy

In reduced or low molecular oxygen environments, facultative anaerobes can use facultative respiration to shift their metabolic pathways and use nitrate NO_3^- as the terminal electron acceptor. The process is called denitrification and is generally depicted by:

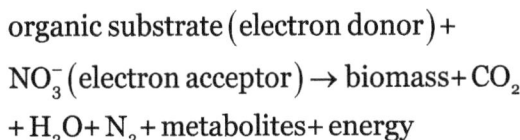

organic substrate (electron donor) +

NO_3^- (electron acceptor) $\rightarrow$ biomass + CO_2

+ H_2O + N_2 + metabolites + energy

The reduction of NO3 – to nitrogen gas N2 is completed through a series of electron transport reactions as follows:

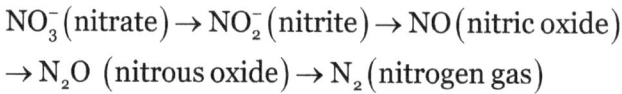

$$NO_3^- \text{(nitrate)} \rightarrow NO_2^- \text{(nitrite)} \rightarrow NO \text{(nitric oxide)}$$
$$\rightarrow N_2O \text{ (nitrous oxide)} \rightarrow N_2 \text{(nitrogen gas)}$$

Most denitrifiers are existing in soil and heterotrophic. A large number of species can reduce nitrate to nitrite in the absence of oxygen, with a smaller number of species can complete the reaction by reducing nitrous oxide to nitrogen gas.

Many classes of microorganisms can survive in the absence of molecular oxygen, using a process called anaerobic respiration, where metals such as Fe^{3+} and manganese Mn^{4+}, sulfate SO_4^{2-}, or even CO_2 can be used in lieu of oxygen to accept electrons from contaminants being degraded. The anaerobic microorganisms that are important to environmental remediation include iron and manganese reducing bacteria, and sulfanogenic and methanogenic bacteria. Anaerobic growth is less efficient than aerobic growth. However, these bacteria complete important geochemical reactions including bacterial corrosion, sulfur cycling, organic decomposition, and methane production. In addition to new cell mass, the byproducts of anaerobic respiration depend on the electron acceptor, and may include reduced forms of metals, carbon dioxide CO_2, hydrogen sulfide H_2S, and methane CH_4. The simplified biochemical reactions performed by these classes of microorganisms are depicted as follows:

1. Iron reduction:

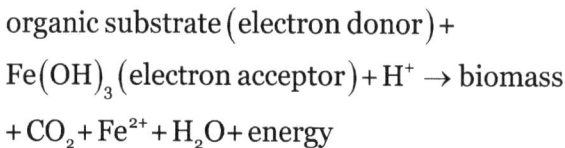

organic substrate (electron donor) +

$Fe(OH)_3$ (electron acceptor) + $H^+ \rightarrow$ biomass

$+ CO_2 + Fe^{2+} + H_2O +$ energy

2. Manganese reduction:

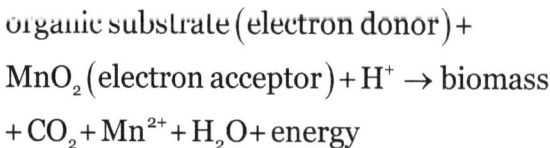

organic substrate (electron donor) +

MnO_2 (electron acceptor) + $H^+ \rightarrow$ biomass

$+ CO_2 + Mn^{2+} + H_2O +$ energy

3. Sulfanogenesis:

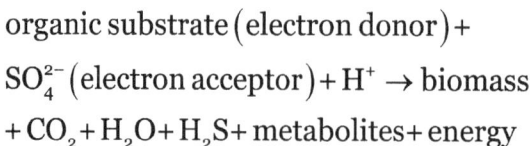

organic substrate (electron donor) +

SO_4^{2-} (electron acceptor) + $H^+ \rightarrow$ biomass

$+ CO_2 + H_2O + H_2S +$ metabolites + energy

4. Methanogenesis:

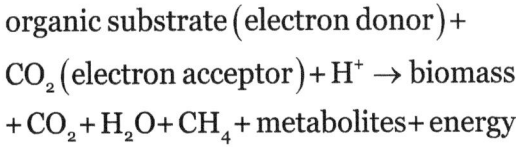

$$\text{organic substrate (electron donor)} +$$
$$CO_2 \text{ (electron acceptor)} + H^+ \rightarrow \text{biomass}$$
$$+ CO_2 + H_2O + CH_4 + \text{metabolites} + \text{energy}$$

Fermentation is a metabolism that can play an important role in oxygen-free environments, where organic contaminants serve as both electron donor and electron acceptor. Through a series of internal electron transfers catalyzed by microorganisms, organic contaminants are converted to innocuous fermentation products, such as acetate, propionate, ethanol, hydrogen, and carbon dioxide. These products can be biodegraded by other species of bacteria to carbon dioxide, methane, and water ultimately. In some cases, microorganisms can transform contaminants with little or no benefit to the cell by the process of cometabolism. For example, when microorganisms oxidize methane, toluene, and phenol, they produce certain enzymes that decompose chlorinated solvents even though the solvents itself cannot support microbial growth. Methane, toluene, and phenol are the primary substrates as they are the microorganisms' primary food sources, while the chlorinated solvents are the secondary substrates. Microorganisms can detoxify halogenated organics contaminants by the process of reductive dehalogenation. Microorganisms catalyze a reaction that replaces a halogen atom on the contaminant molecule by a hydrogen atom, and thus adds two electrons to the contaminant molecule to reduce it. A substance other than the halogenated contaminant, such as hydrogen and low molecular weight organic compounds including lactate, acetate, methanol, or glucose, must exist to serve as the electron donor. In most cases, the reductive dehalogenation process cannot generate energy but is an incidental reaction that may benefit the cell by eliminating a toxic material.

Regardless of the metabolism mechanism that microorganisms use to decompose contaminants, the elemental cellular components of microorganisms are relatively constant as tabulated in Table. If any of these or other elements essential to cell building is in short supply relatively to the carbon content provided by organic contaminants, microbial growth may be limited and rate of bioremediation may be retarded. Therefore, a bioremediation system must be properly designed to supply appropriate concentrations of these nutrients if the natural habitat does not supply them adequately.

Table: Molecular composition of a bacterial cell.

Element	Dry mass by proportion (%)
Carbon	50
Oxygen	20
Nitrogen	14

Hydrogen	8
Phosphorus	3
Sulfur	1
Potassium	1
Sodium	1
Calcium	0.5
Magnesium	0.5
Chlorine	0.5
Iron	0.2
Others	0.3

When the natural environmental conditions at the contaminated site can adequately provide all the essential materials, intrinsic bioremediation can occur in-situ without human participation. However, in-situ engineered bioremediation may be required to supply microorganism-stimulating materials and to optimize the environmental conditions so as to accelerate the desired biodegradation reactions.

The favorable environmental conditions do not always exist and cannot be developed in the subsurface, for example, the site temperature is too cold or the soil is too dense. At these sites, the contaminated soil has to be dug up for cleanup above ground where heaters and soil mixing can help improve environmental conditions. The proper nutrients and oxygen can be added by stirring the mixture or by forced air circulation. However, some microorganisms work better under anaerobic conditions. Mixing soil can sometimes cause harmful contaminants to evaporate before the microorganisms can decompose them. Under these circumstances, the contaminated soil has to be mixed inside a special tank or building where evaporating contaminants can be collected and treated, so as to prevent them from polluting the air.

Existing microorganisms can detoxify many contaminants. However, some compounds, such as petroleum hydrocarbons, alcohols, ketones, and esters, are more easily biodegraded than others. Technologies to stimulate the growth of microorganisms to degrade a wide range of other contaminants, such as PAHs, ethers, halogenated compounds, PCBs, and nitroaromatics, are emerging. Reactions and products of biodegradation of these compounds are given by Alexander. Microorganisms cannot destroy metals, but they can change their reactivity and mobility. Schemes using microorganisms to mobilize metals from one location and scavenge tem from another location have been applied to remediate sites contaminated by mining activities.

Microorganims can also demobilize contaminants by three basic ways: (1) Microbial biomass can sorb hydrophobic organic molecules. Sufficient biomass grown in the path of contaminant migration may stop or slow contaminant transport in the form of a biocurtain. (2) Microorganisms can produce reduced or oxidized species that cause metals to precipitate. Examples are oxidation of Fe^{2+} to Fe^{3+}, which precipitates as ferric hydroxide $FeOH3$; reduction of SO_4^{2-} to sulfide S, which precipitates with Fe^{2+} as pyrite FeS, or with mercury Hg^{2+} as mercuric sulfide HgS; reduction of hexavalent chromium Cr^{6+} to trivalent chromium Cr^{3+}, which can precipitate as chromium oxides, sulfides, or phosphates; or reduction of soluble uranium to insoluble U^{4+}, which precipitates as uraninite UO_2. (3) Microorganisms can biodegrade organic compounds that bind with metals to keep the metals solubilized. Unbound metals often precipitate and become immobilized.

Phytoremediation

Phytoremediation is the removal, stabilization, and degradation of organic and inorganic contaminants in soils by green plants as shown in figure. Water and nutrients are taken up by plants, and carbon dioxide, oxygen, water, and photosynthates are released to the environment. Utilizing plants to control soil and water degradation has a long history. Many early agriculturalists developed plantbased systems to minimize soil erosion, restore disturbed environments, and cleanse water.

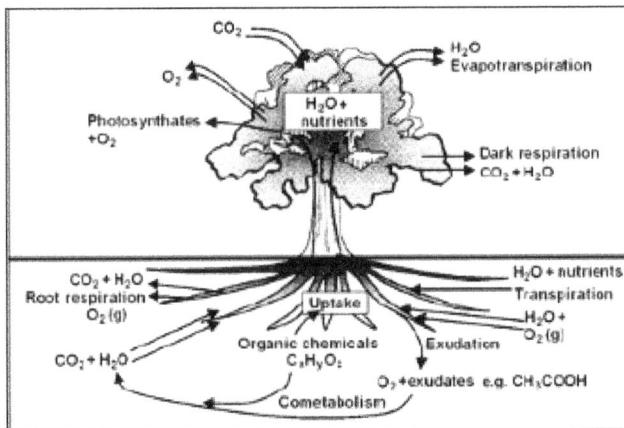

Concept of phytoremediation.

Phytodegradation, also known as phytotransformation, is the breakdown of contaminants taken up by plants through metabolic processes within the plant, or the breakdown of contaminants external to the plant through the effect of constituents, such as exudates and enzymes, produced by the plants. Organic contaminant molecules are degraded into simpler molecules and incorporated into plant tissues by various biochemical reactions caused by enzymes within the plant.

Rhizodegradation, also known as enhanced rhizosphere biodegradation, phytostimulation, or plantedassisted bioremediation/degradation, is the breakdown of contaminants in the soil through microbial activities enhanced by the rhizosphere, i.e., the root zone. It is much slower than phytodegradation. Microorganisms can breakdown organic contaminants such as fuels or solvents into harmless products by biodegradation.

Exudates released by plant roots provide nutrients for microorganisms to enhance their activities. Biodegradation is also aided by plants through loosening the soil and transporting water to the root zones.

Phytovolatilization is the uptake and transpiration of contaminants by plants, with release of the contaminants or modified forms of the contaminants to the atmosphere. It occurs as plants take up water and organic contaminants to the leaves and evaporate, or volatilize, them into the atmosphere.

Phytoextraction, also known as phytoaccumulation, is the uptake and translocation of metal contaminants in the soil by plant roots into the above ground portions of the plants. Certain plants, called hyper-accumulators, absorb unusually large amounts of metals. After these plants have grown for some time, they are harvested and then either incinerated or composted to recycle the metals. If plants are incinerated, the residual ash must be disposed of in a hazardous waste landfill, but the volume of ash will be less than 10% of the original volume of the contaminated soil. Metals such as nickel, zinc, and copper are the best candidates for removal by phytoextraction as it has been shown that they are preferred by a majority of the approximately 400 known hyper-accumulators.

Rhizofiltration is the adsorption or precipitation onto plant root surfaces, or absorption into the roots of contaminants that are in the soil solution in the root zone. It is similar to phytoextraction, but the plants are used primarily to remediate contaminated groundwater rather than soil. The plants to be used are raised in greenhouses with their roots in water. Once a large root system has been developed, contaminated water is used to substitute their water source so as to acclimate the plants. The plants are then transplanted in the contaminated zone where the roots take up the contaminated water. As the roots become saturated with contaminants, they are harvested. For example, sunflowers were used successfully to remove radioactive contaminants from pond water in a test at Chernobyl, Ukraine.

Phytostabilization is the use of certain plant species to immobilize contaminants in the soil and groundwater through absorption and accumulation by roots, adsorption onto root surfaces, or precipitation within the rhizosphere. The process reduces the mobility of contaminants to minimize contaminant migration to the groundwater or air, so as to reduce bioavailability for their entry into the food chain. The technique can be used to reestablish a vegetative cover at sites where natural vegetation is lacking due to high metals concentrations in surface soils or physical disturbances to surficial materials. Metal-tolerant species can be used to restore vegetation to the sites, thereby decreasing the potential migration of contaminants through wind erosion and transport of exposed surface soils, and leaching of contaminants to groundwater.

The processes occurring within the rhizosphere are integral to phytoremediation. Contaminants in contact with plant roots in the root zone must permeate the root membranes by the process of rhizofiltration before they can be absorbed by plants. Uptake of contaminants by plant roots is a direct function of contaminant concentrations in the soil solution and usually involves chemical partitioning on root surfaces followed by transport across the cortex to the plants' vascular systems. The contaminant may be bound or metabolized at any point during transport.

The technology is effective for the remediation of soils contaminated with moderately hydrophobic organic contaminants such as BTEX, chlorinated solvents, and nitro-toluene ammunition wastes with $1.0 \leq \log K_{ow} \leq 3.5$. The octanol-water partition coefficient K_{ow} of an organic compound is the concentration of the organic compound dissolved into octanol divided by the concentration dissolved into water when the organic compound is shaken with a mixture of n-octanol and water. Generally, constituents with log $K_{ow} < 0.5$ are too water soluble to be taken up into roots, and constituents with log $K_{ow} > 3$ are bound too tightly to soil particles or roots to be taken up into plants. Examples of organic compounds with log $K_{ow} < 0.5$ include methyl tertiary butyl ether (MTBE) and4-dioxane, and constituents with log Kow > 3 include most PAHs. Organic contaminants taken up by plants can be degraded by phytodegradation, accumulated in the plant tissue by phytoaccumulation, or transpired through the leaves by phytovolatilization. Phytodegradation of organic contaminants continues in root zones through the process of rhizodegradation. The fate of organic contaminants can generally be predicted using the octanol-water partition coefficient K_{ow} of the particular constituent using Brigg's Law as tabulated in table.

Table: Fate of organics predicted by Briggs Law.

Log K_{ow}	Mechanisms
< 1.0	Possible uptake & transformation
1.0 - 3.5	Uptake, transformation, volatilization
> 3.5	Rhizosphere bioremediation or phytostabilization

The technology is also effective for the remediation of soils contaminated with excess nutrients such as nitrate, ammonium, or phosphate; and heavy metals. Inorganic contaminants can be incorporated in plant tissues through the process of phytoaccumulation. The tendency of plants to uptake, immobilize, or exclude metals is highly contaminant- and soil-specific. Soil factors that influence the tendencies include: (1) soil pH - increases in soil pH generally reduce solubility of metals and uptake by plants; (2) cation exchange capacity (CEC) - increases in CEC of soil reduces plant uptake; (3) organic matter - inorganic forms of metals are generally taken up more readily by plants than organic forms; and (4) natural and synthetic complexing agents - the presence of complexing agents such as ethylenediaminetetraacetic acid (EDTA) and diethylenetriaminepentaacetic acid (DPTA) generally increases the solubility of metals, making them more available to roots and more likely to be taken up by and accumulated in plants.

Metals can also be bound to soil within the rhizosphere, or to the root tissue itself. Exudates released in the rhizosphere can increase the soil oxygen content and the soil pH up to 1.5 pH units, change the redox conditions of the soil, promote oxidation of metal contaminants, and reduce mobility and bioavailability of metals.

Phytoremediation technologies offer these advantages: (1) it is an in-situ technology; (2) it is relatively inexpensive as it uses plants; and (3) it is a safe and passive technology driven by solar energy, and eyepleasing, hence it is more likely to be accepted by the public. However, they have these disadvantages: (1) relatively shallow cleaning depths, less than 1 m for grasses, less than 3 m for shrubs, and less than 6 m for deep-rooting trees; (2) the process is slow and it requires three to five growing seasons to achieve remediation goals; (3) knowledge to optimize phytoremediation technologies is still experimental; (4) the technology is site specific and the choice of plants is critical, and the cleanup strategy requires detailed site characterization to maximize plant growth and contaminant uptake; (5) potential contamination of the food chain; and (6) the technology may relocates contaminants from the subsurface to the plant, creating residual waste to be disposed of.

Although many mechanisms of phytoremediation are not fully understood, there are many reported applications of phytoremediation in China, Portugal, Russia, India, New Zealand, and Australia.

Air Sparging/Soil Vapor Extraction

Air sparging is an in-situ process where air is bubbled through contaminated soil via air injection wells (sparge points) to create a subsurface air stripping system to remove volatile contaminants through volatilization as shown in figure. In a typical field setup, compressed air is delivered to an array of air injection wells through a manifold system to inject the air into the subsurface below the known lowest point of the contaminated zone. The injected air rises toward the ground surface through the contaminated zone due to buoyancy, and the contaminants are partitioned into the vapor phase. The contaminant-laden air eventually reaches the unsaturated vadose zone where the contaminated air is collected by a soil vapor extraction system for further treatment above ground.

Soil vapor extraction by itself is a technology for removal of VOCs and SVOCs from unsaturated soils or the vadose zone. A vacuum is applied to the contaminated soil through vapor extraction wells to generate a negative pressure gradient in the subsurface that drives the vapor to move towards the wells by advection. The contaminant-laden vapors extracted from the wells are then treated above ground using standard air treatment techniques such as carbon filters or combustion.

Different processes including volatilization, diffusion, advection, and desorption occur during soil vapor extraction. Volatilization of contaminants occurs as the contaminated air in soil pores is removed. When the concentration of the volatile organic compound in the pore air is decreased, the shift in chemical equilibrium drives the dissolved and free-phase contaminants to partition from the saturated zone into the unsaturated vadose zone for removal by the soil vapor extraction system.

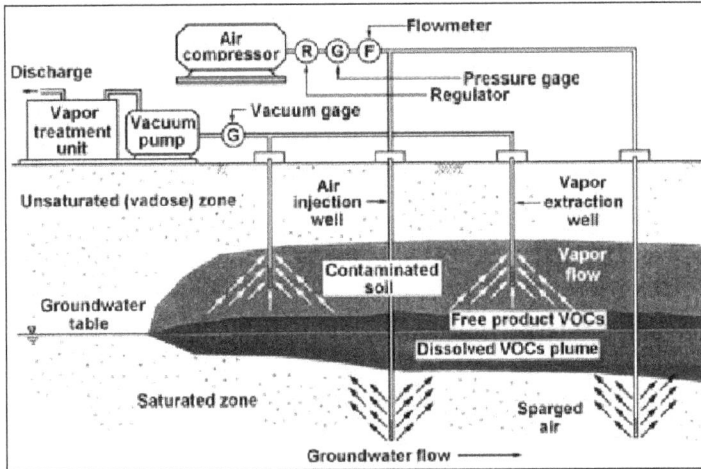

Schematic of in-situ air sparging/soil vapor extraction system.

Advantages of the air sparging technology include: (1) ability to remove VOCs and SVOCs from groundwater including less volatile and sorbed contaminants that are not amenable to soil vapor extraction alone; (2) enhancement of biodegradation by injection of oxygen; (3) enhancement of co-metabolism of chlorinated organics when methane is added to the groundwater; and (4) most effective for relatively permeable and homogeneous sites. Limitations of the technology include: (1) very limited effectiveness in fine-grained and low hydraulic conductivity soils; (2) site geology and depth to contaminants must be known; (3) potential for uncontrolled flow of dangerous vapors through the saturated zone; (4) potential for free product migration from groundwater mounding; and (5) may be ineffective if air flow does not reach the contaminated zone due to soil heterogeneities.

Advantages of the soil vapor extraction technology include: (1) equipment is readily available and easy to install; (2) minimum disturbance to the site; (3) short treatment duration ranging from six months to two years under optimal conditions; (4) economical relative to other remediation technologies; (5) effective for treating both dissolved and free-phase (free-product) contaminants; and (6) very easy to couple with other remediation technologies. However, disadvantages of the technology include: (1) contaminant removal is very rapid at the beginning, and then lingering contaminants are found to exist for a prolonged period of time; (2) it is ineffective in low hydraulic conductivity soils, stratified soils, and high-humic-content soils; (3) air emission treatment systems and permits are often required; and (4) it is applicable to unsaturated soils only.

The soil vapor extraction process can be enhanced by the use of hot air injection, electric heating, or pneumatic fracturing of soil, etc. to increase the mobility of volatile organics and improve the efficiency of the soil vapor extraction process.

Electrochemical Remediation

Contaminants in fine-grained soils cannot be efficiently or effectively removed by the pump-and-treat or soil flushing technology because of the low hydraulic conductivity and large specific area of the soil. Too low a hydraulic gradient applied to fine-grained soils will take too long to complete the cleanup process, as the rate of permeation of flushing fluid through the soil is too low. Too high an applied hydraulic gradient may induce hydraulic fractures in the soil. These fractures may provide paths for contaminants to spread randomly in the subsurface and aggravate the situation. In hydraulically heterogeneous fine-grained soils, the flushing fluid permeates contaminated soils preferentially through paths of the least hydraulic resistance, rendering zones of low hydraulic conductivity practically untreated during the flushing process. The large specific area of fine-grained soil further complicates the situation by providing numerous active reaction sites for soil-contaminant interactions such as surface complexation and sorption/desorption of contaminants. Electrochemical remediation may be a viable remediation solution under these given complex site conditions.

The electrochemical remediation processes rely heavily on the electrokinetic phenomena in fine-grained soils, which stem from the molecular structure of clay. The surfaces of clay particles are normally negatively charged because of isomorphous substitutions and the presence of broken bonds. Adsorbed cations are held tightly to clay particle surfaces by strong electrostatic forces to maintain electrical neutrality. Surplus cations and their associated anions exist as salt precipitates. When clay particles are in contact with water, the salt precipitates go into solution. As the adsorbed cations generate a much higher concentration near clay particle surfaces, they are driven by a concentration gradient to diffuse away from clay particle surfaces to homogenize the ion concentration in the pore fluid. The escaping tendency is counteracted by the electrical attraction of the negatively charged clay particle surfaces. A diffuse double layer is thus formed. Governing equations describing the behavior of the diffuse double layer are given by Hunter, Sposito, Yeung, and Mitchell.

The diffuse double layer provides a mobile layer of cations in the vicinity of the soil particle surface. However, the electrokinetic behavior and thickness of the layer depends heavily on the chemistry of pore fluid. If the interconnecting soil pores are idealized as a bundle of capillaries, the wall surface of the capillary is negatively charged and the mobile cations form a concentric shell in close proximity of the wall surface within the capillary. The behavior of the shell depends on the electrokinetic properties of clay particle surfaces, chemistry of the pore fluid, and their interactions. Several electrokinetic phenomena thus arise in clay when there are couplings between hydraulic and electrical driving forces and flows. The electrokinetic phenomena of direct relevance to electrochemical remediation are electroosmosis, electromigration or ionic migration, and electrophoresis, in which the liquid, dissolved phase, and solid phase moves relative to each other under the influence of an externally applied direct-current (dc) electrical field.

Electroosmosis is the advective movement of the pore fluid from the anode (positive electrode) towards the cathode when a dc electric field is imposed on a wet clay. Electromigration or ionic migration is the advective movement of the charge-carrying dissolved phase relative to that of the liquid phase under the influence of an imposed dc electric field. Anionic species (negatively charged ion) moves towards the anode (positive electrode) and cationic species (positively charged ion) towards the cathode (negative electrode). Electrophoresis is the advective transport of charged particles, colloids, or bacteria in suspension under the influence of an imposed dc electric field.

Electrochemical remediation technologies of contam inated soils may include electrochemical mobilization of contaminants, electrokinetic extraction of contaminants from fine-grained soils, electrochemical injection of decontamination agents, electrochemical transformation of contaminants in soils and their various combinations and modifications. All these remediation technologies involve basically the application of a direct-current (dc) electric field across the soil and utilization of the resulting electrokinetic flow processes, geochemical processes, and electrochemical reactions to remediate the contaminated soils. These electrokinetic flow processes, geochemical processes, and electrochemical reactions include flow of electric current through the soil; electroosmotic migration of pore fluid; electromigration of ions, charged particles, colloids and bacteria; electrolysis of pore water in soil at the electrodes and subsequent migration of hydrogen and hydroxide ions into the soil, resulting in a spatial and temporal change of soil pH; gas generation at electrodes; development of non-uniform electric field; occurrence of reverse electroosmotic flow; changes in electrokinetic properties of soil; hydrolysis; phase change of contaminants; soil-contaminant interactions such as sorption and desorption of contaminants onto and from soil particle surfaces; formation of complexes of contaminants; precipitation of contaminants etc., and interactions of these processes. The migration of pore fluid, ions, charged particles, colloids, and bacteria can be utilized to remove contaminants from polluted soil and/or to inject enhancement agents, nutrients, etc. to facilitate various remediation processes. The geochemical processes can be used to provide the necessary environmental conditions to control the direction of electroosmotic flow and to solubilize contaminants in the soil, so as to enhance the efficiency of the electrochemical remediation processes.

Electrokinetic extraction is an emerging technology developed to remove inorganic and organic contaminants from fine-grained soils as an electrical gradient is a much more effective force in driving fluid flow through fine-grained soils. It involves the application of a dc electric field through electrodes embedded in the subsurface across the contaminated soil. The contaminant is removed by the combination of (1) electroosmotic advection of the pore fluid flushing the contaminants; (2) electromigration or ionic migration of contaminants that carry charges; and (3) electrophoresis of charged particles and colloids that are binding contaminants on their surfaces as shown in figure. Moreover, these chemical transport mechanisms can be utilized to inject cleansing fluid, enhancement agents such as complexing agents and surfactants, nutrients, and/or bacteria to improve the effectiveness and efficiency of the process.

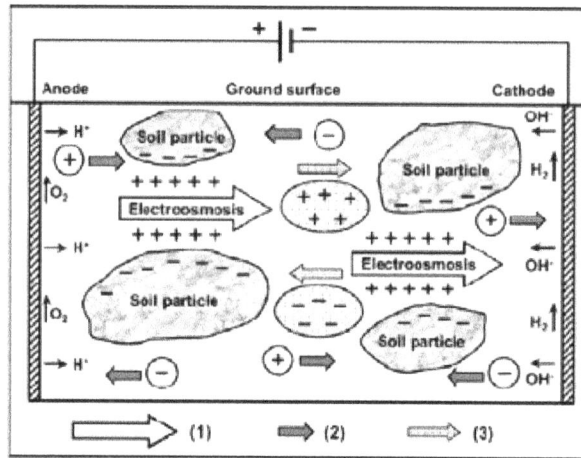

Concept of electrokinetic extraction of contaminants from soil.

Extractability of contaminants from soil primarily depends on the mobility of contaminants within the soil matrix, which is a function of the chemical state of the contaminants, surface characteristics of soil particles, chemistry of pore fluid, and their interactions. In addition to electroosmotic advection, electrokinetic extraction also makes use of the transport mechanisms of advective electromigration and electrophoresis. Therefore, the contaminants must exist as a mobile phase within the soil matrix, such as a dissolved phase in the pore fluid, a colloidal phase suspended in the pore fluid, and/or a mobile immiscible liquid phase co-existing with the pore fluid in soil pores. It is difficult to remove contaminants from a soil that exist as a separate solid phase such as precipitates in soil pores, or as a sorbed phase on soil particle surfaces. Therefore, electrochemical reactions associated with electrokinetic extraction that would affect the mobility of contaminants are of paramount importance on the extractability of contaminants. Factors affecting contaminant extractability by electrokinetics includes: (1) soil type; (2) contaminants type and concentrations; (3) soil pH; (4) acid/base buffer capacity of soil; (5) zeta potential of soil; (6) electroosmotic flow direction; (7) sorption/desorption characteristics of soil particle surfaces; (8) operating parameters; and (9) enhancement techniques.

The pertinent geochemical processes affecting the electrochemical remediation processes include: (1) generation of pH gradient; (2) change of zeta potential of soil particle surfaces; (3) change of direction of electroosmotic flow; (4) sorption and desorption of contaminants onto/from soil particle surfaces; (5) buffer capacity of soil; (6) complexation; (7) oxidation-reduction (redox) reactions; and (8) interactions of these processes.

Electrokinetic extraction is an emerging remediation technology applicable to fine-grained soils of low hydraulic conductivity and large specific area. As with many other remediation technologies, electrokinetic extraction has its own drawbacks including: (1) migration of contaminants is not highly selective; (2) acidification of soils to promote mobility of contaminants may not be technically feasible and/or environmentally acceptable; (3) the technology is not very cost-effective when the target contaminant concentration is low and the background non-target ion concentration is high; among many others. Successful application of the technology primarily depends on mobility of the contaminant in the soil matrix. However, the efficiency and effectiveness of electrokinetic extraction can be improved by combining the technique with other remediation technologies such as the Fenton treatment process, Lasagna process, biodegradation, phytoremediation process, etc.

Ek-fenton Process

The Fenton reaction involves two steps: (1) decomposition of H_2O_2 catalyzed by Fe^{2+} or other transition elements to generate hydroxide radicals; and (2) oxidation of organic pollutants by hydroxide radicals. Yang and Long, and Yang and Liu studied the feasibility of coupling electrokinetic extraction with a Fenton-like treatment process using a permeable reactive wall of scrap iron powder to remove and oxidize organic contaminants experimentally. Their bench-scale laboratory experimental results indicate that it is feasible to combine electrokinetic extraction and the Fenton-like process to treat TCE and phenols in soils. The overall contaminant remediation efficiency is contributed by two mechanisms: (1) organic contaminant destruction by the Fenton-like process; and (2) contaminant removal by electrokinetic extraction.

Lasagna Process

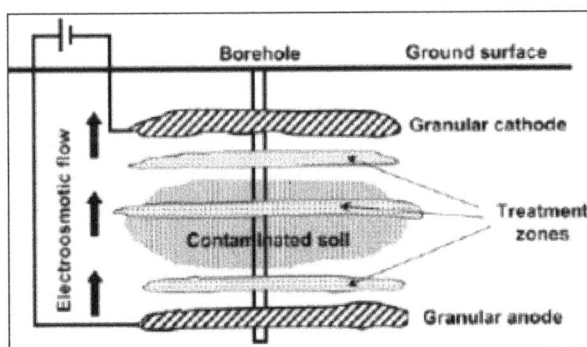

Horizontal configuration.

The Lasagna process is an Integrated In situ Remediation technique. Electrokinetics is coupled with sorption/degradation of contaminants in treatment zones that are installed directly in contaminated soils. A dc electric field is applied to mobilize the contaminants in contaminated soils to the treatment zones where the contaminants are removed by sorption, immobilization, or degradation as shown in figure. The technique is called Lasagna because of the layered appearance of electrodes and treatment zones. Conceptually, it can treat organic and inorganic contaminants as well as mixed wastes.

Electrodes and treatment zones can be of any orientation depending upon the emplacement technology used and the site-contaminant characteristics as shown in figure.

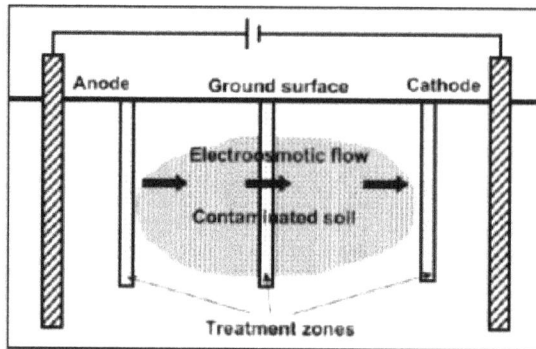

Vertical configuration Principle of Lasagna Process.

Generation of in-situ Reactive Iron-rich Barriers

Faulkner et al. have successfully generated subsurface barriers of continuous iron-rich precipitates in-situ by electrokinetics in their laboratory-scale experiments. Continuous vertical and horizontal iron-rich bands up to 2 cm thick have been generated by applying an electrical voltage of less than 5 V over a period of 300-500 hours, with an electrode separation of between 15 and 30 cm. The iron-rich barrier is composed of amorphous iron, goethite, lepidocrocite, maghemite, and native iron. The thickness of the iron-rich band increases with an increase in the applied voltage. The barrier is generated using sacrificial iron electrodes installed on either side of a soil/sediment mass as shown in figure. The applied dc electric field dissolves the sacrificial anode and injects the iron ions into the system, and reprecipitation of the iron ions in an alkaline environment forms the barrier. The iron-rich band so produced is of hydraulic conductivity of $1 \times^9$ m/s or less and unconfined compressive strength of 10.8 N/mm^2. The barrier may function as an impervious barrier to contaminant transport or a reactive barrier to degrade contaminants. By monitoring the dc electric current intensity passing through the barrier, the integrity of the iron-rich band may be assessed. Moreover, the barrier may 'self-heal' by continuing application of a dc electric current.

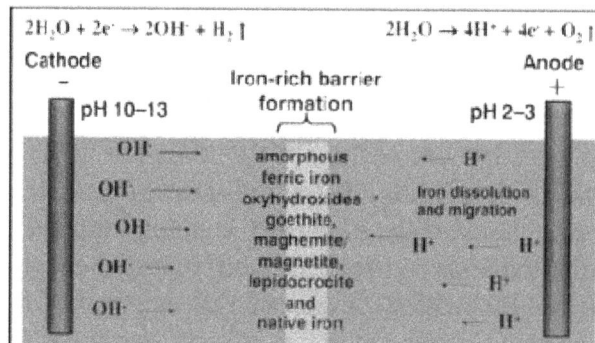

Principle of iron-rich barrier generation by electrokinetics.

Bioelectrokinetic Remediation

The feasibility of injecting benzoic acid to enhance the biodegradation of TCE by electroosmosis for neutral benzoic acid and electromigration for benzoate anion was explored by Rabbi et al. Their experimental results demonstrate the feasibility in principle for using electrokinetic injection to engineer the degradation of recalcitrant hydrocarbons, or other difficult to degrade contaminants. A similar study on phenol was conducted by Luo et al.

Bioleaching is a process for remediation of metalcontaminated soil. Indigenous sulfur-oxidizing bacteria convert reduced sulfur compounds to sulfuric acid which acidifies the contaminated soil and mobilize the metal ions. Experimental results on remediation of coppercontaminated soil by Maini et al. indicate that the effectiveness of electrokinetic extraction was enhanced by pre-acidification of the contaminated soil by sulfuroxidizing bacteria. The electrokinetic treatment also appeared to stimulate the activity of sulfur-oxidizing bacteria by the removal of inhibitory ions and other positive effects of the electric current upon soil microbial activities. The synergistic methodology appears to be promising for a range of contaminated sites including former gasworks and wastes from mining.

Ek-phytoremediation Process

The use of a combination of electrokinetic remediation and phytoremediation to decontaminate soils contaminated by copper, cadmium, and arsenic was investigated by O'Connor et al. in laboratory-scale reactors. Their results indicate that a dc electric field can migrate metallic contaminants from the anode towards the cathode, accompanied by significant changes in soil pH. Moreover, perennial ryegrass could be grown in the treated soils to take up a proportion of the mobilized metals into its shoot system.

Other possible uses of electrokinetics in environmental management include: (1) concentration, dewatering and consolidation of wastewater sludge, slimes, coal washeries, mine tailings, or dredging; (2) injection of grout to control groundwater flow; (3) injection of cleaning agents into contaminated soils ; (4) injection of vital nutrients to stimulate the growth of microorganisms for biodegradation of specific contaminants; (5) electrokinetic barriers to contaminant transport through compacted clay landfill liners of slurry encapsulation walls; (6) electrokinetic extraction of contaminants from polluted soil; (7) modification of flow pattern of groundwater, and manipulation of the movement and size of contaminant plumes; (8) enhancement of flow through permeable reactive contaminant barriers; (9) real-time and reliable detection and delineation of subsurface contamination; (10) rapid and reliable in-situ determination of hydraulic conductivity of compacted clay landfill liners; (11) in-situ generation of reactants for cleanup, electrolytic decomposition and/or solidification of contaminants; and (12) retrofitting of leaking in-service geomembrane liner.

Thermal Treatment

Thermal treatment technologies are destruction and removal types of heat treatment of contaminated soil including thermal desorption, vitrification, thermal destruction, and incineration.

Thermal desorption is a technology that heats contaminated soil or sludge in-situ or ex-situ to volatize contaminants and remove them from the soil. Volatile and semi-volatile organics are removed from contaminated soil in thermal desorbers at 100 to 300°C for lowtemperature thermal desorption, or at 300 to 550 °C for high-temperature thermal desorption. The vapors are collected and treated in a gas treatment system. The gas treatment units can be condensers or carbon adsorption units that trap organic compounds for subsequent treatment or disposal. The units can also be afterburners or catalytic oxidizers that completely destroy the organic contaminants. The basic components of a thermal desorption system include: (1) pretreatment and feed; (2) thermal processor and discharge; and (3) air emissions control. Critical factors controlling the performance of the technology are: (1) soil type and moisture content; and (2) temperature, residence time, mixing, and the sweep gas flow rate in the thermal processor. The technology is effective in the remediation of soils contaminated with oil refining wastes, coal tar wastes, wood-treating wastes, creosotes, hydrocarbons, chlorinated solvents, fuels, PCBs, mixed wastes, synthetic rubber processing waste, pesticides, and paint wastes.

Vitrification is a thermal stabilization/solidification process that does not require the addition of reagents. It can be performed in-situ or ex-situ. The process involves melting and fusion of materials at temperatures normally in excess of 1200°C followed by rapid cooling to glasslike materials. During the process, non-volatile metals are immobilized within the glass; volatile metals, such as lead, cadmium, and zinc, volatilize, and they must be captured in the off-gas treatment system; and organic compounds pyrolyze or combust. The process requires an extraordinary amount of energy of 800 to000 kWh per ton of soil treated in-situ. As a result, it is only used to treat relatively small quantities of wastes that are difficult to treat by other remediation technologies. It can be used to treat radioactive wastes, metal sludges, asbestos-containing waste, or soil or ash contaminated with metals. Although soil or waste containing organic contaminants can be treated by the process, vitrification technologies are usually directed towards inorganic contaminants. Although the gas structure is generally a relatively strong and durable material resistant to leaching, proper disposal of the vitrified slag is required. It should be noted that vitrification is a complex process and requires a high degree of specialized skill and training.

Thermal destruction is ex-situ processes that thermally destroy organic contaminants by oxidation, pyrolysis, hydrogenation, and reduction.

Catalytic oxidation has been used for emission control of organic compounds from industrial sources and/or remediation of contaminated sites for a long time. However, oxidation catalysts have been poisoned by halogens, certain metals, particulates, phosphorus, and sulfur compounds in the past, thus limiting their applications to non-halogenated contaminated air streams. Recently, new catalysts have been developed to overcome these obstacles and to achieve high destruction efficiencies.

Catalytic oxidation units for the destruction of halogenated compounds typically consist of: (1) a gas or electric preheater to raise the air-stream temperature to the catalyst temperature of < 450 °C; (2) a catalytic reaction; (3) a shell and tube heat exchanger to recover approximately 50% of the heat in the reactor exit gas; and (4) a scrubber to remove halogens and hydrogen halides from oxidation products before their release to the atmosphere, as shown in figure. In an air sparging/soil vapor extraction operation, the contaminated vapors are extracted from wells through a manifold and pumped to the catalytic oxidation unit. In a groundwater remediation operation, the contaminants are air stripped from the groundwater and the contaminated vapors are directed to the catalytic oxidation unit.

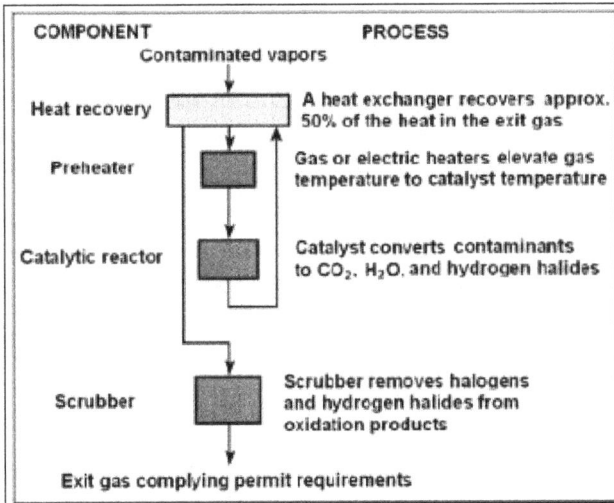

Catalytic oxidation unit flow chart.

Catalytic oxidation is cost competitive with other processes as the fuel costs are lower. However, a potential disadvantage of the technology is that the catalyst may be deactivated or poisoned by various volatile chemicals. Nonetheless, it is anticipated that development of new catalysts will expand the range of halogenated organic compounds that can be effectively treated by catalytic oxidation.

Pyrolysis is an ex-situ chemical decomposition process by which the wastes are heated in the absence of oxygen as shown in figure. As it is impossible to achieve a completely oxygen-free environment in practice, nominal oxidation occurs. Pyrolysis involves a twostep process to remediate contaminated soil in a chamber at temperatures 400 to 1200 °C: (1) wastes are heated to separate the volatile components from the non-volatile char and ash; and (2) the volatile components are burnt to assure incineration of all the hazardous components.

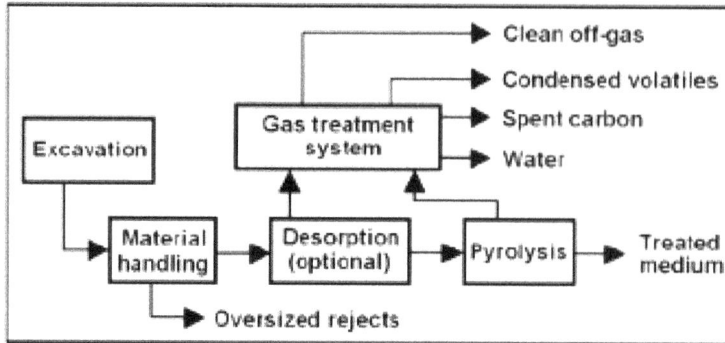

Schematic of pyrolysis.

Incineration is intended to permanently destroy organic contaminants. It is a complex system of interacting pieces of equipment, representing an integrated system of components for waste preparation, feeding, combustion, and emissions control as shown in figure. The most important component of the system is the combustion chamber, or the incinerator. There are four major types of incinerator: (1) rotary kiln; (2) fluidized bed; (3) liquid injection, and (4) infrared.

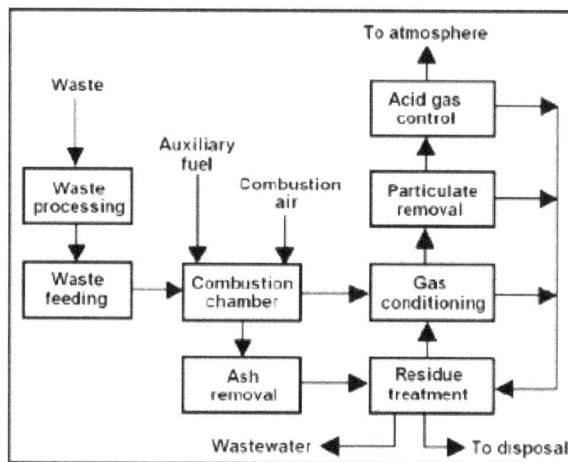

Incineration system concept flow diagram.

Performance Monitoring

Performance monitoring is used to ensure that the behavior of the contaminant plume does not deteriorate over time and/or the remedial action is progressing as intended. It collects a subset of parameters used in the site characterization monitoring focusing on the most significant parameters of the site as a function time during the remediation process. Performance monitoring wells should be located up-gradient from, within, transverse to, and just down-gradient from the contaminated zone. Specimens collected from these wells are used to verify that the concentrations of individual chemicals or parameters of concern, contaminated zone boundaries, and overall progress towards remediation goals are acceptable over time and space.

The minimum goals of performance monitoring of remediation processes are to: (1) demonstrate that remediation is occurring as expected; (2) detect changes in subsurface environmental conditions that may reduce or enhance the efficiency of the remediation technology; (3) verify that the contamination situation is not deteriorating; (4) verify that there has been no unacceptable impact to down-gradient receptors; (5) detect any new releases of contaminants to the environment that may pose an unacceptable risk to public health and the environment or impact the effectiveness of the remediation technology; (6) demonstrate the efficacy of institutional controls already in place to protect the public health and the environment; and (7) verify progress towards achievement of remediation objectives. In additional to meeting these goals, a site-specific contingency plan must be prepared in case the remediation technology fails to perform as anticipated.

On completion of the remediation process, a verification sampling and analysis program is typically implemented to confirm that cleanup goals have been met. The verification data are compared with the final remediation goals using statistical methods. Additional remediation may be necessary if the remediation goals are not met.

Selection of Remediation Technologies

A contaminated site is an unusable resource. However, the site can be restored for productive use with adequate site remediation planning and proper executive of remediation technologies. Therefore, upon identification of a contaminated site, these planning questions should be posed to address the remediation aspects of the site: (1) Under which law or statute should the remediation be conducted? (2) What is the extent of contamination and what is the likelihood that the contaminants will migrate? (3) If remediation is necessary, what level of cleanup is required to restore the site to be a clean site? (4) What remediation technologies are available and which options are the preferred technologies for the site?

The applicability and effectiveness of a remediation technology to treat a particular site depends on four major factors: (1) the local factor: the acceptability of the technology to local community; (2) the regulatory factor: the statutes and regulations applicable to the operation of the technology, for example, laws that regulate discharges to surface waters and emissions to the atmosphere, as discharges and/or emissions may be necessary for the operation of the technology; (3) the technology factor: the technical, cost, and performance characteristics of the technology; and (4) the site factor: the hydrogeology of the contaminated site; and the structure, particle size distribution, organic content, pH, and moisture content of the soil at the site.

Evaluation of remediation technologies may be in three phases: (1) general response actions; (2) assembly of technologies as alternatives; and (3) screening of alternatives. A general response action is defined as one approach to remediation of contaminants in one medium. It can be active remediation or institutional actions. Remediation alternatives to be assessed are scenarios for the total remedial action. Each may include several individual remedial actions for different parts of the contaminated site and different media. Each alternative scenario may include several general response actions and many remediation technologies.

If a technology is considered to be applicable for the contaminated site, the implementation feasibility at the particular site can be evaluated by analyzing whether the technology can achieve: (1) overall protection of public health and the environment; (2) compliance with applicable or relevant and appropriate requirements; (3) longterm effectiveness and permanence; (4) reduction of toxicity, mobility, and volume of contaminants; (5) shortterm effectiveness; (6) no insurmountable implementation barriers; (7) relatively cost-effective; (8) compliance with legal requirements; and (9) community acceptance.

A screening matrix using 13 factors was developed to evaluate qualitatively the performance of different remediation technologies by the U.S. Environmental Protection Agency. Eight of these 13 factors involve a comparative rating: better, average, or worse. The eight comparative rating factors are: (1) overall cost including the design, construction, and operation and maintenance costs of the main processes of the technology; (2) commercial availability; (3) minimum contaminant concentration achievable by the technology; (4) time to complete remediation of a "standard" site of200 tons for soil or785,000 liters for groundwater using the technology; (5) degree of system reliability and maintainability when using the technology; (6) level of awareness of the remediation consulting community regarding the availability of the technology; (7) degree of regulatory and permitting availability of the technology; and (8) level of community acceptability to the technology. The remaining five factors demonstrate performance-related questions. The five technology performance factors are: (1) Is the technology capital intensive or operation and maintenance intensive? (2) Is the technology typically part of a treatment train? (3) What is the physical form of the residuals generated when the technology is applied? Is it solid, liquid, or vapor? (4) What parameter of the contaminated media is the technology primarily designed to address? (5) What is the long-term effectiveness of the technology? Does the use of the technology maintain the protection of public health and the environment over time after the remediation objectives have been met? The matrix provides a screening of prospective technologies in a rapid way.

Metal-contaminated Soil Remediation

Phytoremediation

Phytoremediation refers to the technologies that use living plants including herbs (e.g. Thlaspi caerulescens, Brassica juncea, Helianthus annuus) and woody (e.g. Salix spp., Populus spp.) species, to clean up soil, air, and water contaminated with hazardous contaminants using their ability to either contain, remove, uptake, or render harmless various environmental contaminants like potentially-toxic elements, organic compounds and radioactive compounds in soil or water, thanks to their transport capacity and accumulation of contaminants. The use of plants for in situ treatment of contaminated soils was suggested for first time in the early 1990s. The term phytoremediation was then introduced early in the same year to describe the use of plants for extracting PTE from soils. Phytoremediation can be applied to inorganic as well as organic contaminants. As stated by, plants are kind of "chemical factories" that exercise great influence on their environment not only by uptake of substances but also by exudation of many molecules that are produced in primary and secondary metabolism. This lively chemical and physical interaction of plants with their environment are of great utility often use for the remediation of contaminated sites; refers to as phytoremediation.

The successful application of phytoremediation techniques is dependent on many parameters among which, contaminants must be bioavailable and ready to be absorbed by roots. The bioavailability of metals depends from solubility of the metals in soil. Nevertheless, mechanisms and efficiency of the phytoremediation depend not only on the bioavailability of metals but also on several others factors such as the nature of contaminant, soil properties, and plant species. The plants which are generally considered for this purpose are those that exhibit great efficiency in phytoremediation processes. They are commonly named as "hyperaccumulator", macrophytes capable of tolerating and accumulating metals present in the soil ≥ 10 g kg^{-1} (1%) Mn or Zn, ≥ 1 g kg^{-1} (0.1%) As, Co, Cr, Cu, Ni, Pb, Sb, Se or Tl, and ≥ 0.1 g kg^{-1} (0.01%) Cd of the dry mass of shoots on soils rich in PTE in the aerial organs from soils without suffering phytotoxic damage; while yielding low biomass.

Otherwise, the extraction efficiency of the pollutants also depends on the biomass produced by the plant. Indeed, the bigger is the biomass the higher the ability of the plant to uptake big quantity of metals. However, more harvests, time and effort will be required to remove the plants after treatment. This will determine the total cost of the entire operation, including disposal, incineration or composting of biomass. Phytoremediation is a reliable reclaiming treatment, because it does not interfere with the ecosystem, it requires less manpower and therefore cost-effective compared to traditional physicochemical methods. Phytoremediation techniques could be applied for the recovery of the industrial sites heavily contaminated with low to moderate concentration.

Mechanisms of Phytoremediation

The removal of inorganic pollutants and even organic using phytoremediation is made possible following diverse mechanisms summarized in the figure.

1. Phytoextraction: Metals are extracted from the soil by the plant and transferred to the plant's shoot and leaves. Plants which are often used in this process are selected based on their ability to accumulate contaminants and produce a high biomass.

2. Phytoimmobilization/Phytostabilization: In this process, pollutants are absorbed and immobilized in the root system and it is reduces their mobility. It has been used for the removal of Pb, As, Cd, Cr, Cu and Zn.

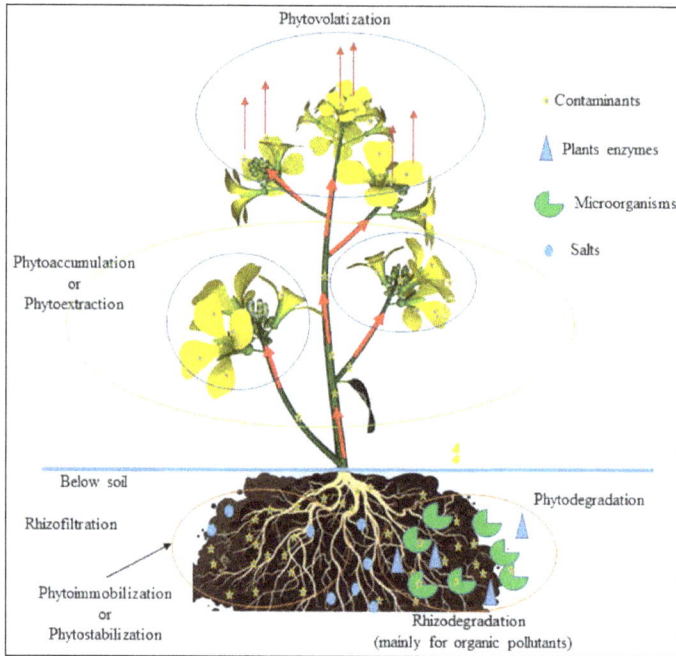

Different mechanisms involve in phytotechnology.

3. Phytovolatilization: Pollutants are absorbed at root level and converted in a less toxic forms as a result of metabolic modification and released in atmosphere from the aerial parts of plant. We can thus state that this mechanism only relocate the pollutants from the soil to the air. However, in anyway, the soil has been sanitized.

4. Phytodegradation: This mechanism is mainly for the sequestration of organic contaminants in the soil. It involves Plant enzymes to degrade organic contaminants. Various enzymes are involve in the mechanism among which: (i) dehalogenase (sequestration of chlorinated compounds); (ii) peroxidase (sequestration of phenolic compounds); (iii) nitroreductase (sequestration of explosives and other nitrate compounds); (iv) nitrilase (sequestration of cyanated aromatic compounds); (v) phosphatase (transformation of organophosphate pesticides). At this level, phyto and bioremediation cannot be separated from one another, as microorganisms play an important role in these phytotechnologies. In fact, plants are in continuous interaction with microorganisms, some of which form close associations or symbiotic relationships. This phenomena is what explain the symbiosis that form mycorrhizal fungi with almost all land plants and nitrogen-fixing rhizobia with legumes.

5. Rhizofiltration: This mechanism is commonly applied for the removal of pollutants from surface water or wastewater through adsorption or precipitation on the roots. It has been used for metals and even radioactive elements removal from soil, wastewater

and contaminated water with satisfactory results. This technique requires the adjustment of the pH of the medium a better efficiency of the operation; this is seen as a disadvantage of the technique.

6. Rhizodegradation: Just like phytodegradation, this mechanism permit to degradation of organic pollutants in the rhizosphere through rhizospheric microorganisms. It involves a continuous interaction between plants and microorganisms; and thus it cannot be separated from bioremediation.

7. Phytodesalination: This technique is really not used for remediation of contaminated-coil with PTE or persistent organic pollutants but used for the removal of slat from salt-affected soil; it is made possible using halophyte plants (Artemisia argyi, Limonium bicolor, Melilotus suaveolens and Salsola collina). Halophytes are plants with great ability to tolerate high concentrations of Na^+ and Cl^- ions; making them able to reclaim excessive saline soil. To be noticed, it is reported that saline soils cover about 6% of the world's land and it well known that salinity is the main environmental factor limiting plant growth and productivity.

Advantage and Disadvantage of Phytoremediation

In comparison to many other remediation technologies, phytoremediation is found to be of low costs, it protects the soil from erosion (reduction of erosion rate), improves the chemical, physical and biological soil properties, and enhances land esthetic. Phytoremediation is a technology that meets consensus and is highly accepted by the population. It is suitable for sites with low to moderate contamination and where contaminants diffused over large areas, and where there are no temporal limits to the intervention, and finally, it requires less human power. However, despite all this advantages, phytoremediation presents also some limitations which are worth to be mentioned. Indeed, it is time consuming, strong dependence upon: climatic conditions, contaminants concentration and bioavailability, plant tolerance to contaminants, contamination area extent and depth (limited by the rhizosphere or the root zone). The disposable of harvested wastes is another challenge of phytoremediation. It is also not suitable for severely contaminated site such as e-waste contaminated site where potentially-toxic elements and persistent co-exist (the growth of plant would be inhibited), it is also not suitable when arable land (usable land for agricultural production is limited). Therefore, at this stage, another technology would be needed to tackle the remediation of the site. For a better performance of phytoremediation, it could also be combined to electrochemical process. However, the challenge is that the combination would somehow inhibit some phytoremediation processes such as phytodegradation, rhizodegradation which only take place with continuous soil's microorganisms. Indeed, the electrochemical process which includes the induction of low level direct current in the soil via electrodes would provoke the rising of soil's temperature and the change of soil pH; and thus disturb or inhibit the activity of bacteria. As a consequence, the performance of plant to remove the contaminants will be affected.

Chemical Leaching

Chemical Leaching and Leaching Agents

Chemical leaching is one of the traditional remediation technologies used for contaminated soil remediation; and it involves dissolution, extraction and separation of the pollutants. Chemical leaching is one of the common and widely used methods for soil and sludge's PTE removal. Through the precipitation, ions exchange, chelation or adsorption, the PTE in soil are transferred from soil to liquid phase, and then separated from the leachate. The separated pollutants are then converted to the appropriate form before disposal or can be reinserted in the recycling circle. For the dissolution and extraction process, there must be a step of breaking the bound between metals and soil constituents. The success this operation requires the use of acids, oxidants and complexants. Originally, contaminated soil is treated with strong inorganic acids such as HCl, HNO_3, H_2SO_4, H_3PO_4. Unfortunately, the application of the above-strong acids have been found to be environment and ecological disastrous. Indeed, strong acids have a strong capacity of destroying soil structure, and killing soil's microorganisms. Otherwise, in the process of sanitizing the soil using strong acids, there also occurs the loss of soil constituent which is of great concern for the ecological consideration. Such situation is not in line with the protection of the environment on one hand, and does inhibit the productivity of the treated soil on the other hand. As a consequence, the use of strong acids is not environmental friendly. Thus, the integrated utilization of acids or reagents should be deliberately selected to fulfill the requirement of target contaminants removal on one hand, and soil ecological protection on the other hand. This justifies the introduction of Low molecular weight organic acids such as acetic acid, oxalic acid, which constitute a group of weak organic acids and chelating reagents such as nitrilotriacetic acid (NTA), sodium tripolyphosphates (STPP) and ethylenediaminetetraacetic acid (EDTA). The use of weak acids showed mitigated results even though promising. On the other hand, chelating agents develop great affinity with the metals ions and possess prominent properties of oxidizing and forming complexes with metals cations; which could improve their extraction efficient. The use of organic chelators has been widely investigated and results are satisfactory; mainly EDTA is well known for its excellent ability to recover metals from soil (80%) depending on the type of soil. However, these chelators seem to be refractory to the environment and not easily biodegradable and thus can pose a secondary pollution via leaching to the groundwater. As a consequence, there is a need to find more suitable chelators for the replacement of the refractory ones. In line with this objective, the use of organic acids and new generation of chelating agents are increasingly been investigated as an alternatives to above-mentioned washing reagents. N, N-bis(carboxymethyl) glutamic acid (GLDA), a chelator with excellent biodegradability, more than 60% degradable within 28 days. According to the OECD 301D test with lowest 'eco-footprint' characteristics in comparison to EDTA and STPP; has been suggested due to it exceptional chelating capacity towards different divalent metal ions. It was successfully used by and for the recovery of Cd Co, Cr, Cu, Ni and Zn from dewatered sewage sludge. The removal efficacy was comprised between 60 and 86% and 94% for both studies, respectively. In addition, it comparison with citric acid during the work of showed great efficacy and efficiency of GLDA compared to citric acid. The more a chelators possesses a carboxyl group (-COOH), the higher its performance would be during soil washing process. It is only used in an ex-situ remediation technology, which create too much disturbance of soil system and its microorganisms. Here below are some organic chelators used in soil washing technology.

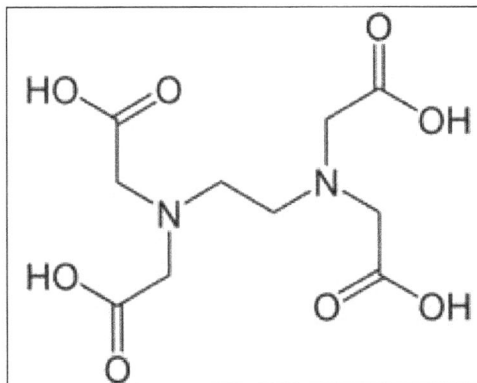

Some organic chelators often used for soil washing, EDTA and NTA are commonly used, while others in the table are known as new generation of chelators.

Challenges Related to Field Application of Chemical Leaching

During chemical leaching, the use of significant amount of chelating agent is essential for the mobilization of PTE within the soil system. The addition of chelants to soils not only promotes metals mobilization and transfer from the soil to the chelants' solutions but it also increases the total concentration of the soluble metals. A better mobilization of metals in the soil requires up to hundreds of mill molar per liter concentration of the chelating agents in the soil solution. The issue is that the process can recover only part of the concentration of the dissolved metals, and leaching will be unavoidable; which could lead to the possible contamination of the ground water and slow (several weeks or months) decomposition of the synthetic organic acids. Following the application of chelate forming agents, the removal of metals may continue for a long time. Besides, the use of chelating agents could exercises adverse effects on the soil microorganisms.

Otherwise, except the fact that during the soil washing/leaching process, soil minerals and other constituents are washing away together with the target pollutants, the in situ application of this technology at the large scale would be very challenging. Indeed, the injection of washing reagent in the soil is really challenging as it would not be easy to control the flow direction; and the solution will tend to flow vertically (leaching towards ground water) rather than in the desired direction, generally horizontal. As a consequence, the in situ field applicability of the technology at the large scale is limited; only ex-situ applications are widely known. Otherwise, the technology is solvent consuming and involve longue processes and post treatments of the treatment waste and thus time consuming with high requirement of human power. Otherwise, it is soil generate too much soil disturbance (soil returning). One of the alternatives to make valuable this technology is to combine it with other technology which permits the control of the solvent flow with less soil disturbance such as electrochemical process. This combination has given birth to the electrokinetic remediation technology.

Electrokinetic Remediation

Principles and Mechanisms of Inorganic Contaminants Removal in soil

Electrokinetic remediation is a technique that consists in displacing or moving pollutants in contaminated soil from their contaminated points towards a specific controlled extraction points which are generally the electrodes cells. This technique is made possible by the application of a direct low current between electrodes well-disposed in the soil in order to optimize the electric field. The principle of pollutants cleanup is controlled by some key processes such as electroosmosis, electromigration and electrophoresis. These mechanisms involve different mechanism. Electroosmosis knows as electroosmotic flow, consists of the displacement of the liquid in the porous soil as result of the application of the electric field. During this movement, the pore fluid carries along organics and neutral molecules. Electromigration consists of the transport of charged particles (anions and cations) towards the opposite electrode cell. As for the electrophoresis, it is the movement of dispersed particles in the medium relative to a fluid as result of a spatially uniform electric field. These mechanisms are of great importance in pollution remediation (soil and sediment treatment) when using electrokinetic approach.

During electrokinetic remediation, there occur electrochemical reactions of which, electrolysis of water represents one of the most important and influential reactions. These reactions take place on the surface of the electrodes as the result of the application of low direct electric current. During electrolysis process, there occur a generation of protons (H+)$H+$ on the anodic surface and hydroxyl ions (OH−)$OH−$ on the cathodic surface; which lead to an important pH gradient. These ionic species are mobilized through the soil at a rate determined mainly by the electromigration and diffusive processes and the soil's buffering capacity.

The pH profile is a key parameter during soil treatment with electrokinetic approach. Indeed, the changes of pH induce beside electrokinetic processes, physicochemical processes among which precipitation/dissolution of minerals and metals, adsorption/desorption of pollutants and ion exchange between the soil solid and the pore water. As it is well known, pH exercises strong influence on the chemical speciation of the compounds mainly inorganic present in the soil system. It determines the state or ionic forms in which a compound is found in the soil. This will indirectly condition the predominant transport mechanism by which this compound will move during the treatment.

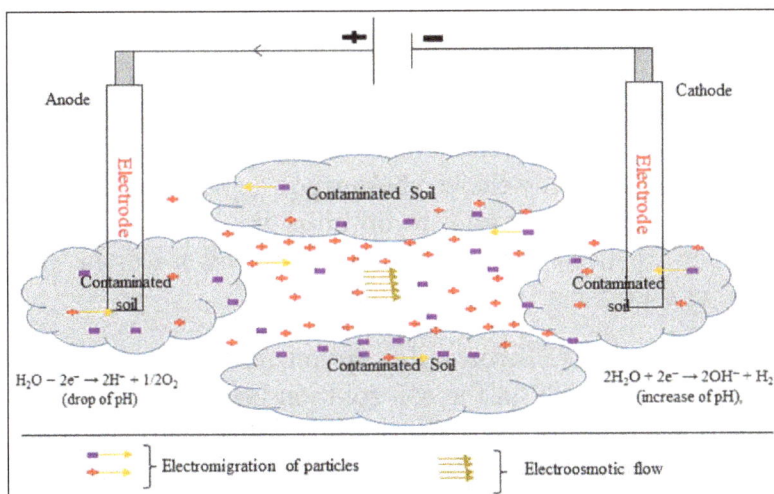

Mechanism of electrokinetic remediation approach.

Especially the change in pH affects the surface charge of soil particles and metal ions mobility. The generated acidic condition helps mobilize sorbed metal ions, prevents formation of metal hydroxide and carbonates precipitates; and thus facilitates their electromigration via the electroosmotic flow of the liquid. However, highly acidic conditions cause electroosmotic flow to stop or reverse, whereas alkaline condition results in PTE precipitation and increases electroosmotic flow. Thus, to maintain this parameter within a suitable range, pH control if often performed in both anode and cathode by adding sodium hydroxide (0.1 and 1 M) and acetic acid/citric acid (0.1 and 1 M) respectively. The in-situ acidification, however, may not be adequate if the soil possesses high buffering capacity. Moreover, the generated base front causes metal ions to precipitate, impeding their final arrival at the cathode. Consequently, external/artificial acidification is often required even necessary during electrokinetic soil remediation. However, the use of strong inorganic acids such as HCl, HNO_3 is not is not recommended as it can damage the soil structure. In addition, it would be costly and is not environmentally acceptable. Generally, water or chemical solutions [(0.1 M) EDTA or acetic acid, citric acid, etc.] are continuously injected at the anode to maintain optimal remediation conditions; contaminated water is removed at the cathode by pumping.

This technology has been successfully used in single for the treatment of various wastes/sites such as wastewater, sewage sludge, soil and sediments contaminated with inorganic and organic pollutants. However, to optimize its efficacy, it has also been used in the combinations with other technologies. The combination of electrokinetic remediation method with other technologies has been tested and is still on the hotspot of scientific research in environmental filed. It includes electrokinetic-microbe joint remediation, electrokinetic-chemical joint remediation, electrokinetic-oxidation/reduction joint remediation, coupled electrokinetic-phytoremediation, electrokinetics coupled with electrospun polyacrylonitrile nanofiber membrane, and electrokinetic re-mediation conjugated with permeable reactive barrier.

Electrodes and Electrolytes

Various inert electrodes made of ceramic, carbon, graphite, titanium, stainless steel, are generally used during electrokinetic remediation of contaminated-soil. Each electrode has its level of stability; the choice of electrode depends on the use and purpose. The electrodes are configured in order to optimize the electrical field in the treated area. Generally, they are disposed in the contaminated soil at 1.1.5 m spacing, with imposed DC current at 1.3.0 V cm^{-1} or 500 kWh m^{-3}.

Electrokinetic extraction of PTE involves desorption/dissolution followed by transport. When the concentration of PTE in the soil solution becomes below the soil sorption capacity, chemical additives are typically needed to help mobilize and sorb metals. Also poor conductivity-pollutants (in the form of sulfides) or present in metallic form (Hg) cleanup involve a primary step of dissolution. This step generally involves the use of some appropriate electrolytes such as distilled water, organic acids or synthetic chelates; which aims to enhance the efficiency of the remediation. Several chemical have been tested as additives and include acetic acid (CH_3COOH), citric acid (($HOOC$-$CH_2)_2C(OH)(COOH)$)), nitrilotriacetic acid (NTA), ethylene-diamine-tetra-acetic acid (EDTA), ethylenediaminedisuccinic acid (EDDS), diethylenetriaminepentaacetic acid (DTPA), and potassium iodide (KI). These additives also known as enhancement fluids mobilization efficiencies varies from one to another and depending on the type of metal species in soil. It is worth to mention that the removal efficiency varies not only depending on the type of the chemical used (anolyte) and metal remediated but also on the type of electrode. Indeed, the use of KH_2PO_4 as an anolyte permitted to enhance the removal efficiencies of as species by >50% and ~ 20% for Cu species. Meanwhile, it did not enhanced the removal of the Pb and Zn (< 20%). Also reported that adding ethylene diamine disuccinate (EDDS) in the anolyte enhanced Pb and Cd removal efficiencies in the contaminated soil.

Advantages and Limitations

Electrokinetic technology has many advantages among which, it applicability for in-situ/ex-situ remediation, applicable to low-permeability soils and a mixture of contaminants where other technologies cannot be applied, applicable to a wide range of pollutants, and applicable to heavy and severely contaminated sites. However, the main limiting factor for direct electrokinetic remediation is the fluctuation in soil pH; because it cannot maintain soil pH value. Therefore there is a need to control the soil pH by external intervention through the addition of buffer solutions in cathode and anode cells. In fact, controlling the pH in the electrode cells remains the main challenge of this technology. Electrokinetic remediation has shown promising results and is still under development stage.

Soil Bioremediation

Anthropogenic activities such as industrial, mining and military processes are the major sources that contributed to widespread contamination of the environment throughout the world with numerous chemicals including petroleum hydrocarbons, polyaromatic hydrocarbons (PAHs), polychlorinated biphenyls (PCBs), halogenated dibenzodioxins/furans, chlorinated solvents, pesticides and toxic heavy metal(loid)s. Consequently, several thousands of sites around the world are seriously polluted requiring remediation. The costs for cleaning up of contaminated sites are extremely high, and in the USA alone about $8 billion are spent annually. Global costs are in the range of 500 billion. Traditional methods for remediation of contaminated soils include dig and dump, excavation, transport, landfilling, soil washing, the addition of oxidants (hydrogen peroxide or potassium permanganate) and incineration. Due to the high cost of remediation technologies, several polluted commercial properties were abandoned or idled rather than remediated. There are over 500 000 of these so-called brownfields in the USA with an estimated clean-up and redevelopment costs more than $650 million. Almost 800 000 potential brownfield sites have been identified in Europe.

The land is scarce which supports life on earth and soil is not a renewable resource. Cleaning up of contaminated soil and its protection are key priorities for redeveloping land and urban regeneration in developed or industrialized countries. Industrialization together with technological advancements over the past more than 60 years has led to the creation of large areas of abandoned or underused and potentially contaminated lands in cities and suburbs throughout the world, and these are classified as brownfield sites. As the cities grew outwards, brownfields became located in the centre of cities often occupying high-value lands. Brownfield sites pose a risk to human and environmental health have negative impacts on the economy at the regional level by becoming obstacles for urban development; therefore, cleaning up of these sites have become priorities for many nations. According to the USEPA brownfield site is "real property, the expansion, redevelopment or reuse of which may be complicated by the presence or potential presence of a hazardous substance, pollutant or contaminant". Brownfields contain co-contaminants. For the past decade, there has been an increasing awareness and interest among the public for sustainability in remediation, especially in the developed countries. Sustainable remediation not only brings great opportunities but also challenges, for both researchers and the practitioners in the remediation area. Sustainability considers that the resources are finite and should be used judiciously to meet the needs of current but without compromising the future generations. Thus, the benefits of sustainable remediation are realized through the promotion of renewable energy, material recycling, preservation of natural resources and minimization of waste and energy. The traditional physicochemical technologies for soil remediation cannot be considered as sustainable because these technologies do not include the criteria for sustainability. Over the past decade, green and sustainable remediation is gaining importance as a beneficial approach to optimize all phases of remediation. Bioremediation mediated by biological agents such as microorganisms (bacteria, fungi, algae, etc.) or plants is considered a cost effective, green and sustainable approach for restoring the contaminated sites. However, bioremediation has its limitations for its field-scale application as an efficient remediation technology other than for petroleum hydrocarbon contaminated sites.

Bioremediation Approaches

Bioremediation approaches can be applied either in situ or ex situ depending on the nature of contaminant and site conditions. In situ treatment is more attractive and cost effective as it is not or less disruptive and does not involve excavation and transport of contaminated soils. The commonly used in situ approaches include natural attenuation, biostimulation, bioventing and bioaugmentation. In contrast, the ex situ approaches involve excavation and removal of contaminated soil for treatment either on the site or transportation to a suitable place before treatment. The commonly used ex situ bioremediation approaches include land farming, biopiles and bioslurries. Each contaminated site or brownfield represents a challenge due to its former use and depending on whether it is abandoned or underused, and the contamination is real or perceived. Biotechnological interventions are required to bring back these sites to their beneficial uses. Bioremediation approaches when combined with sustainable practices such as the use of renewable sources (e.g., solar or wind power instead of fossil fuel based energy or generation of biomass for bioenergy) will result in greater environmental, economical and societal benefits.

Natural Attenuation

Natural attenuation processes involve contaminant attenuation to harmless products through natural processes, such as microbial degradation, volatilization, sorption and immobilization. The natural attenuation process is contaminant specific and commonly employed for petroleum hydrocarbon contaminated sites. However, natural attenuation may not be a suitable option for several other contaminants such as persistent organic pollutants. Although natural attenuation has proven to be a successful approach to treat petroleum contaminants (benzene, toluene, ethylbenzene and xylene), it may not work if the site does not have the contaminant degrading microorganisms or nutrients.

Biostimulation

The microbial transformation of contaminants in soils depends on the availability of nutrients (carbon, nitrogen, phosphorus and potassium), favourable environmental conditions (pH, electrical conductivity, aeration, temperature) and the nature of contaminant itself and its bioavailability. Some contaminants such as persistent organic pollutants (e.g., PAHs, PCBs, lindane, dichlorodiphynyltrichloroethane) are extremely insoluble in water and tend to strongly sorb to organic matter in soils thereby decreasing their availability to microbes. The use of biosurfactants can enhance the bioavailability of such pollutants. The addition of slow release fertilizers or organic waste and manures can supply the nutrients and stimulate the indigenous microbes to transform the contaminants.

The addition of natural organic substrates such as mulch and manure has shown to remove perchlorate through stimulation of anaerobic degradation by microbes. Perchlorate reducing bacteria are ubiquitous, have the ability to reduce perchlorate to chloride under anaerobic conditions using perchlorate as a terminal electron acceptor for growth and energy in the presence of electron donor. The bioremediation process using glycerine-diammonium phosphate (DAP) successfully treated over000 tonnes of soil from a 1000 acre Bermite site from Los Angeles, California containing 0.8.4 mg perchlorate/kg soil to non-detectable levels within seven month period, which is considered to be a safe and economical treatment. The former Bermite site was used to manufacture various explosives and related products including perchlorate during 1987.

Composting

The addition of compost or composting is considered to be one of the most cost-effective approaches to remediate contaminated soils because it can increase soil organic matter content and soil fertility besides enhancing bioremediation. Several studies have demonstrated the effectiveness of composting as a technology to detoxify or stabilize a wide range of contaminants including toxic metals, PAHs and pesticides. Sorption of organic contaminants to soil organic matter can decrease the fraction of contaminant that is available to microorganisms for degradation. However, water extractable organic matter from cow manure compost was shown to increase the solubility of certain PAHs phenanthrene, pyrene and benzo-a-pyrene with 8.4, 34 and 89 times higher than their measured concentrations in water, respectively, which enhanced their biodegradation. The observed increase in PAH solubility and biodegradation was attributed to the high molecular weight (>1000 Da) fraction of water extractable organic matter from cow manure. In another study, Wu et al. demonstrated the enhanced bioavailability and removal of PAHs up to 90% in soils contaminated with diesel, coal tar and coal ash when amended with compost. Both degradation and desorption processes were attributed as reasons for the observed PAH disappearance. Degradation of organic contaminants in soil is often difficult due to their low bioavailability. The addition of surfactants to soil can increase the bioavailability of some organic pollutants. Co-composting of PAH polluted sediments with green waste in different proportions for nine months has resulted in a decrease of PAH concentrations to < 1 mg g^{-1}. The co-composted product is considered to have the potential for use as technosol or plant growth substrate in revegetation of urban areas or brownfields.

Pelaez et al. has successfully demonstrated field-scale bioremediation of 900 m^3 PAH polluted soil from a former chemical factory near Oviedo (Spain) used for manufacture of naphthalene, phenols and other chemicals from coal processing, in a biopile using commercially available fertilizer and surfactants, which resulted in 94.4% decrease in PAH contamination during 161 days. The decrease in PAHs coincided with an increase in indigenous bacteria able to degrade PAHs, with Bacillus and Pseudomonas being abundant bacteria.

Bioaugmentation

Introducing specific microorganisms to decontaminate the soils when indigenous microbes are not efficient is considered a more acceptable approach to remediate the contaminated soils. However, the strains for bioaugmentation should ideally have (i) superior ability to degrade the target contaminants, (ii) easy to cultivate, (iii) fast growth, (iv) tolerance to the high concentration of contaminant and (v) ability to survive in a wide range of environmental conditions/stressors. Bioaugmentation has been proven to be successful for a wide range of pollutants including pesticides such DDT, lindane, endosulfan, pentachlorophenol (PCP), polyaromatic hydrocarbons (PAHs) and total petroleum hydrocarbons. However, predation, competition and toxins in soils can negatively affect the survival of introduced microbes. In such cases, bioaugmentation using immobilized cells in carrier materials or preadapted strains to the problem soil conditions may prove to be advantageous regarding enhancing their survival in soils.

Phytoremediation

The use of plants to remediate contaminated sites has been considered as an in situ cost-effective option alternative to the relatively expensive traditional physicochemical technologies based on excavation, dig and dump. However, phytoremediation did not find wide application, especially for metal contaminated sites, due to potential risks to biota via the metal laden biomass. Phytostabilization rather than phytoaccumulation could be an attractive alternative option for remediation of metal contaminated sites. Phytostabilization involves stabilization/immobilization of contaminants in the soil via binding to the roots or complexation through root exudates, which reduces the bioavailability of contaminants, therefore, reduces the risk to food chain. Two heavy metal (Cu, Pb, Zn) contaminated brownfield sites (a former landfill site and an industrial site used for shipyard, wood impregnation, etc.) have been successfully remediated using phytostabilization through willow plants (Salix Klara and Salix singer). This field trial has demonstrated that phytostabilization of brownfield sites with bioenergy crops can provide environmental benefits by turning these areas into economical and beneficial uses. Plants in association with microbes can be applied to remove the labile/bioavailable pool of inorganic contaminants from a site, remove or degrade organic contaminants, stabilize or immobilize contaminants (phytostabilization/in situ immobilization/Phyto-exclusion).

Aided phytostabilization was applied over a six-year period on a 1 ha site previously used for on-land disposal of Zn, Pb and Cd contaminated sediments at Fresnes-Sur-Escaut in northern France. A basic mineral amendment (Optiscor) was applied to the soil, which was then planted at high density with a commercial cultivar of grass (Deschampsia cespitose). The trial showed stabilization of contaminants with effectively 100% vegetation cover (by reducing soil-human contact via direct soil exposure and dust inhalation) and a reduction in plant-metal uptake and transfer. Metal concentrations in the foliage of cover grass were reduced by 60% for Zn and 20% for Cd. Metal concentrations in biomass were sufficiently low to allow subsequent biomass use as compost. In Austria, in situ immobilization/Phyto-exclusion was applied over a 13-year period at Arnoldstein (South Austria) on arable land impacted by Pb/Zn smelter emissions. Gravel sludge and iron bearing materials (red mud) were applied as soil amendments and Cd excluding cultivars of commercial food crops (barley, maize and potatoes) grown with the aim of reducing contaminant transfer from soil to plants and groundwater. Amendment addition resulted in a significant reduction in the labile contaminant pool (80% Cd; Zn > 90% and Pb > 90%) in the soils. Whereas, the Cd uptake by barley was decreased by > 75% compared to an accumulating cultivar. Uptake of Zn, Cd and Pb into maize silage was reduced by 70%, 60% and 50% respectively. Application of soil amendments (such as lime, red mud, zeolites, cyclonic ashes, iron grits and slags, composts, biochar and other organic amendments) has shown to reduce the bioavailability of a wide range of contaminants while simultaneously contributing to revegetation success and thereby, protecting against offsite movement of contaminants by wind and water.

Thus, phytoremediation has emerged as a promising strategy for in situ removal of a wide variety of contaminants. Plants in association with microbes seem to be more effective for removal/degradation of organic contaminants from impacted soils.

About 40% of plant photosynthates are released as sugars, organic acids and other larger organic compounds into soils, which serve as carbon and energy sources for microbes. The flavonoids and coumarins that are released by plant roots can stimulate the growth and activity of PAH and PCB degrading bacteria. During a 60-week study, about a 73% decrease in total PAHs was observed in planted sediments compared with unplanted sediments which showed only 25% decrease. Phytoremediation over a two-year period decreased the total PAH concentration by 30% which is double the unvegetated highly contaminated site. In a 60-day field trial, 96% of 4,6-trinitrotoulene was removed from a test plot by maize (Zea mays). The disadvantages of phytoremediation are that it is a slow process requiring several years and more crop harvests and the challenge is that there are stressors (variation in temperature, nutrients, precipitation, herbivory, plant pathogens, and competition by weeds) that affect phytoremediation in the field but are not encountered in the greenhouse. A successful strategy for overcoming the challenge of plant stress is to use plant growth promoting bacteria that can lower the level of deleterious ethylene and also enhance germination and plant growth rates under stress conditions, particularly when used in conjunction with contaminant tolerant plants species. Plant growth promoting rhizobacteria can also act as biocontrol agents by suppressing the plant pathogens.

Integrated Approaches

In most cases, single remediation technology may not be effective and requires a combination of technologies. Poor bioavailability of persistent organic pollutants (POPs) in soil often impedes the success of bioremediation as a feasible decontamination approach. Fenton—bioremediation is emerging as a promising integrated approach, which enhances POP removal efficiencies. Fenton oxidation followed by bioremediation could improve the effectiveness of bioremediation of highly contaminated soils. The integrated technology combines rapid and aggressive oxidation by Fenton pre-treatment followed by degradation by microbial activity in the pre-treated soil matrix. Efficiencies ranging from 70% to 98% have been reported for combined bioremediation-Fenton treatment for POP contaminated soils. Fenton oxidation combined with bioremediation enhances PAH removal efficiency in several ways. Kao and Wu developed a combined Fenton pre-treatment and bioremediation method to efficiently degrade3,7,8-tetrachloro dibenzo-p-dioxin (TCDD)-contaminated soils. In this study, Fenton pre-treatment removed 98% TCDD. The advantages of Fenton pre-treatment are (i) decrease in pollutant concentrations to levels that are less toxic to soil biota, (ii) improvement of the bioavailability of parent PAH, (iii) prevention of incomplete mineralization of partially oxidized PAHs by utilizing degrading bacteria and fungi which are commonly found in the environment, (iv) release of oxygen from the H_2O_2 decomposition from Fenton treatment that provides aeration for aerobic biological transformation.

Challenges and Prospects

Large areas of land around the world have been impacted by former industrial and other anthropogenic activities. These include urban brownfields; former mining and resource extraction sites and bringing these back to beneficial uses require site-specific approaches. Although bioremediation is considered environmentally beneficial and sustainable, the process can be slow. Current bioremediation technologies suffer from some limitations, which include the lack of adequate understanding of the contaminant degrading capabilities of microbial communities in the field, low bioavailability of contaminants on spatial and temporal scales and lack of adequate knowledge on metabolic cooperation networks among the microbial consortia/communities.

The restoration of natural functions of some contaminated sites may not be feasible and, hence, the application of the principle of function-directed remediation may be sufficient to minimize the risks of pollutants and bring back the lands to beneficial uses. Integrated approaches such as pre-treatment of highly contaminated soils using chemical oxidants in safe concentrations to soil biota, followed by bioremediation, appear to be a promising technology for some of the intractable pollutants. Also, plant-microbe associations have great potential for their application in remediation of contaminated sites. Bioremediation, although green and environmentally safe, should be combined with renewable resources such as the wind, solar energy, and linked to the generation of biomass for renewable energy resources, all of which make bioremediation a more sustainable technology. Successful adaptation of sustainability in remediation is essential, and a concerted action of academia, government and industry are needed for successful implementation.

Bioremediation Techniques in Polluted Tropical Soils

Ecosystems are regularly confronted with natural environmental variations and disturbances over time and geographic space. A disturbance is any process that removes biomass from a community, such as fire, flood, drought, or predation. Disturbances occur over vastly different ranges in terms of magnitudes as well as distances and time periods and are both the cause and product of natural fluctuations in death rates, species assemblages, and biomass densities within an ecological community. These disturbances create places of renewal where new directions emerge out of the patchwork of natural experimentation and opportunity implying a good measure of ecological resilience is a cornerstone theory in ecosystem. One of such disturbances is pollution which alters ecological balance.

Intense industrial activity and urbanization in recent times, especially in developing countries, have led to serious environmental pollution, resulting in a large number and variety of contaminated sites which became a threat to the local ecosystems. In all these, natural resources such as soils, water, air and vegetation are adversely affected.

Industrial revolution gave birth to environmental pollution which continued till today. It was a revolution that led to the emergence of great factories and consumption of immense quantities of fossil fuels which was associated with an unprecedented rise in air pollution and large volume of industrial chemical discharges. This was added to the growing population with a load of untreated human waste.

Pollution defines the introduction of harmful substances often referred to as contaminants into the natural environment that cause adverse change. The term contamination is in some cases used interchangeably with pollution in environmental chemistry, where the main interest is the harm done on a large scale to humans or to organisms or environments that are important to human beings. Common soil contaminants include chlorinated hydrocarbons, heavy metals such as chromium, cadmium—found in rechargeable batteries, and lead—found in lead paint, aviation fuel and still in some countries, gasoline, zinc, arsenic and benzene. Recycling industrial byproducts into fertilizer may result in the contamination of soils with various metals. Ordinary municipal landfills are the source of many chemical substances entering the soil environment and often reaching groundwater, emanating from the wide variety of refuse.

In the case of the term contamination, it is the presence of a minor and unwanted constituent in a material, in a physical body or in the natural environment. In chemistry, contamination usually refers to a single constituent, but in specialized fields the term can also mean chemical mixtures, even up to the level of cellular materials.

Pollution may take various forms including discharge of deleterious chemical substances on natural substances. Pollution can be point source or nonpoint source pollution.

Sometimes pollution takes the form of harmful energy such as noise, heat or light. Generally speaking, foreign substances and energies which contaminate natural resources are referred to as pollutants. Substances contain some level of impurity; and this may become an issue if the impure chemical is mixed with other chemicals or mixtures and causes additional chemical reactions. Sometimes, the additional chemical reactions are beneficial, in which case the label 'contaminant' is replaced with reactant or catalyst. When additional reactions are detrimental, other terms such as toxin or poison depending on the chemistry involved are used. However, if no remedial action is undertaken, the availability of arable land for cultivation will decrease, because of stricter environmental laws limiting food production on contaminated lands. Inorganic and organic contaminants typically found in urban areas are heavy metals and petroleum-derived products. The presence of both types of contaminants on the same site presents technical and economic challenges for decontamination strategies. There have also been some unusual releases of polychlorinated dibenzodioxins, commonly called dioxins for simplicity.

In many countries, there is paucity of soil information leading to several forms of soil degradations. Except in recent times environmental impact assessments (EIAs) are rarely conducted on natural resources before embarking on major projects. The EIAs are often not backed up with necessary implementation legislations. Mineral exploration and exploitation as well as various construction activities are known to have negative impact on surface and subsurface soils, surface and groundwater, rocks and rocklike minerals, atmospheric resources, vegetation and wildlife.

Available soil data are not problem-solving. Non-use of soil survey data and information has led to soil and soil-related environmental problems such as nutrient depletion, nutrient imbalances, multiple nutrient deficiencies, nutrient toxicity, general decline in soil quality and yield decline. The situation is often aggravated by socioeconomic pressures mainly resulting from poverty and inability to afford relevant inputs of agricultural production. Sound characterization and classification of soils based on quality and proper presentation of such information in user-friendly form is a necessary adjunct in sustained use of soils. Again, soil quality data will go a long way in promoting bio-safety of farm products for both local consumption and their internationalization.

Specifically, some biotechnological methods are suggested for the amelioration of contaminated soils. A good knowledge of status and distribution of polluted soils will go a long way in assisting in the production of land use maps which will facilitate policy and legislations on soil and soil-related natural resources. Land use maps derived from soil survey and land evaluation are useful in soil management as well as in vulnerability and risk assessments. This is true as soil quality problems vary requiring different remediation strategies to overcome.

Remediation deals with the removal of pollutants or contaminants from natural resources. The affected natural resources may include soil, groundwater, surface water sediment, vegetation, rock minerals, wildlife and air. A major aim of remediation is the recovery and general protection of human health and the environment. Sometimes, remediation is done in places intended for redevelopment. Remediation goes with an array of regulatory requirements, and its assessments are based on human health and ecological risks.

Several approaches are used in the remediation of polluted soils, ranging from biological, chemical and engineering techniques. Sometimes, it may require a combination of organic and inorganic strategies. For instance the Neapolitan yellow tuff (NYT) was utilized as a component of an organo-mineral sorbent/exchanger soil conditioner with pellet manure (NYT/PM) to reduce the mobility of Cd and Pb and recover plant performance in heavily polluted soils from illegal dumps near Santa Maria La Fossa (Lower Volturno river basin, Campania Region, southern Italy). Pot experiments were performed by adding the NYT/PM mixture (1:1, w/w) to polluted soil at the rates of 0%, 25%, 50% or 75% (w/w). Wheat (Triticum aestivum) was used as the test plant. The addition of organozeolite NYT/PM mixture significantly reduced the DTPA (diethylene-triamine-pentaacetic acid)-extractable Cd and Pb from 1.01 and 97.5 mg kg^{-1} in the polluted soil, to 0.14 and 11.6 mg kg^{-1}, respectively, in the soil amended with 75% NYT/PM. The best plant response was ob-served in amended soil systems treated with 25% NYT/PM, whereas larger additions induced plant toxicities due to increased soil salinity.

When a soil on site is found to be contaminated to a depth of several metres and construction work needs to get started in a few months' time, soil replacement is the fastest remedy. However, some of the contaminated areas can be restored by combining modern and age-old methods. This is where plants and their microbial partners may enter the picture now and in the future. This because heavy metals in soils with residence times of thousands of years present numerous health dangers to higher organisms. They are also known to decrease plant growth, ground cover and have a negative impact on soil micro flora. There is increasing and widespread interest in the maintenance of soil quality and remediation strategies for management of soils contaminated with trace metals, metalloids or organic pollutants. Heavy metals are deposited in soils by atmospheric input and the use of mineral fertilizers or compost, and sewage sludge disposal. Conventional remediation methods usually involve excavation and removal of contaminated soil layer, physical stabilization and washing of contaminated soils with strong acids or HM chelators. Bioremediation, that is the use of living organisms to manage or remediate polluted soils, is an emerging technology. It is defined as the elimination, attenuation or transformation of polluting or contaminating substances by the use of biological processes.

It is no new discovery that many plant species can grow in soils contaminated by various pollutants. Some species can even sequester or decompose contaminants. Soil and plant microbes help plants survive in harsh conditions.

Bioremediation includes the productive use of biodegradative processes in the elimi-nation or detoxification of pollutants that have found their way into the environment, especially where such pollutants are capable of threatening public health. Some of the methods are ex situ while others are in situ. The ex situ bioremediation techniques in-volve the excavation or removal of soil from ground. A good number of in situ biореme-diation techniques are generally the most desirableoptionsdue to cheapness and fewer disturbances since they provide the services in place avoiding excavation and transport of contaminants. Processes include phytoremediation, phytostabilization, phytotrans-formation, phytoextraction, rhizofiltration and phytoscreening.

Phytoremediation involves the treatment of polluted natural resource through the use of plants that mitigate the problem without the need to excavate the contaminant ma-terial and dispose of it elsewhere. The use of plants in remediation has been growing rapidly in popularity worldwide for the last twenty years or so. Phytoremediation may be defined as use of vegetation to contain, sequester, remove, or degrade organic and inorganic contaminants in soils, sediments, surface water and groundwater. Phytore-mediation is a technology that uses plants to remove contaminants from soil and water. The basic idea that plant can be used for environmental remediation is very old and cannot be traced to any particular source. However, a series of fascinating scientif-ic discoveries combined with an interdisciplinary research approach have allowed the development of this idea into a promising, cost-effective, and environmental friendly technology.

Certain plants and microorganisms are able to precipitate metal compounds in the rhizo- sphere. Efficacy was shown by the use of lead pyromorphite, as phytoremedi-ation may provide an effective means to reduce metal toxicity as well as metal mobil-ity. This is referred to as phytoimmobilisation. Although the application of microbial biotechnology has been successful with petroleumbased constituents, microbial diges-tion has met limited success for widespread residual organic and metals pollutants. Vegetation-based remediation shows potential for accumulating, immobilizing, and transforming a low level of persistent contaminants. We can find five types of phy-toremediation techniques, classified based on the contaminant fate: phytoextraction, phytotransformation, phytostabilization, phytodegradation, rhizofiltration, even if a combination of these can be found in nature.

Phytoremediation consists of reducing or eliminating pollutant concentrations in con-taminated soils, water, or air, with plants. Selected plant species are able to contain, degrade, or eliminate metals, pesticides, solvents, explosives, crude oil and its deriv-atives, and various other contaminants from the media that contain them. Boyd and Javre reported phytoenrichement of soils by Sebertia acuminata in New Caledonia. In phytoremediation, the assumption is that certain plants called hyper accumulators are able to bioaccumulate, degrade, or render harmless contaminants found in natu-ral resources such as soils, water, and air. The maize plant (Zea mays) showed high tolerance towards Cr with negligible concentration in leaves. A plant is said to be a hyperaccumulator if it can concentrate the pollutants in a minimum percentage which varies according to the pollutant involved. More than 1000 mg/kg of dry weight for nickel, copper, cobalt, chromium or lead; or more than 10,000 mg/kg for zinc or manganese are recommended. In addition to this, it is assumed that hyperaccumulating plants can be found thriving under very harsh conditions or under situations that are not ideal for plant growth.

Some plants are able to translocate and accumulate particular types of contaminants. Plants can be used as biosensors of subsurface contamination, thereby allowing investigators to quickly delineate contaminant plumes. Chlorinated solvents have been observed in tree trunks at concentrations related to groundwater concentrations. Phytoscreening often leads to more optimized site investigations and reduce contaminated site cleanup costs. Phytoremediation has become increasingly popular and has been employed at sites with soils contaminated with lead, uranium, and arsenic and it has the advantage that environmental concerns may be treated in situ.

The technology of phytoremediation has been successfully used in the restoration of abandoned metal-mine sites, reducing the impact of sites where polychlorinated biphenyls have been dumped during manufacture and mitigation of on-going coal mine discharges.

There are a range of processes mediated by plants which are useful in soil and soil-related environmental problems. Processes include phytostabilization, phytotransformation,phytoex- traction, rhizofiltration and phytoscreening.

Phytostabilization entails the reduction of the mobility of substances in the environment. This could be done by limiting the leaching of substances from the soil. Its main focus is on longterm stabilization and containment of the pollutant. Plants can reduce wind erosion; or their roots can prevent water erosion, immobilize the pollutants by adsorption or accumulation, and provide a zone around the roots where the pollutant can precipitate and stabilize. Phytostabilization focuses mainly on sequestering pollutants in soil near the roots but not in plant tissues. By this, pollutants become less bioavailable to livestock and wildlife, and human exposure is drastically reduced.

Phytoextraction is the uptake and concentration of substances from the environment into the plant biomass. The use of plants to mine toxicants is called phytomining. Phytoextraction employs metal hyperaccumulator plant species to transport high quantities of metals from soils into the harvestable parts of roots and aboveground shoots. Phy- toextraction is an innovation using higher plants for in situ decontamination of metal-polluted soils, sludges and sediments. Large biomass production and high rates of metal uptake and translocation into the shoot system are critical in achieving reason- able metal extraction rates. Effective phytoextraction requires both plant genetic ability and the development of optimal agronomic management practices. Hyper accumulators are defined as plants that contain in their tissue more than000 mg kg-1 dry weight of Ni, Co, Cu, Cr, Pb, or more than000 mg kg^{-1} dry weight of Zn, or Mn. Hyper accumulation is thought to benefit the plant by means of allelopathy, defence against herbivores, or general pathogen resistance in addition to metal tolerance. In-situ phytoextraction of Ni by a native population of Alyssum murale on an ultramafic site (Albania) have been reported. In the case of phytomining, the use of native flora (including local populations of hyperaccumulators) with limited agronomic practices (extensive phytoextraction) could be an alternative to intensively managed crops. The use of plants in remediation has been growing rapidly in popularity worldwide for the last twenty years or so. In general, this process has been tried more often for extracting heavy metals than for organics the technique of phytoextraction uses plants to remove contaminants from soils, sediments or water into harvestable plant biomass. Such organisms that absorb larger-than-normal amounts of contaminants from the soil are referred to as hyperaccumulators. Examples of hyperaccumulators are Athyrium yokoscense (Japanese false spleenwort), Avena strigosa (Brittle oat), Crotalaria juncea (Sunn hemp), Eichhornia cras- sipes (water hyacinth), Pistia stratiotes (water lettuce). Helianthus annuus (Sunflower), Salix viminalis (Basket willow), Lemna minor (Duckweed), Amaranthus retroflexus (Redroot Amaranth), Glomus intradices (Mycorrhizal fungus), Eragrostis bahiensis (Bahia lovegrass), Cynodon dacvtylon (Bermuda grass), Festuca arundinacea (Tall fescue), Lolium perenne (Perennial ryegrass), Panicum virgatum, (Switchgrass), Phaseolus acutifolius (Tepary beans), Cocos nucifera (Coconut tree), Spirodela polyrhiza (Giant duckweed), Tagetes erecta (African-tall) and Zea mays (Maize).

In phytoremediation, plants absorb contaminants through the root system and store them in the root biomass and/or transport them up into the stems and/or leaves. A living plant may continue to absorb contaminants until it is harvested. Thereafter the process, the cleaned soil can support other vegetation with significant healthfulness.

Some transgenic plants containing genes for bacterial enzymes have been found to be effective hyperaccumulators. Salt-tolerant plants like sugar beets are commonly used for the extraction of sodium chloride in reclaiming soils previously flooded by salt water. Sunflower (Helianthus annuus) is an effective hyperaccumulator in cleaning soils contaminated with arsenic. In general, plants with non-invasive and moisture-tolerant root systems can be planted on the embankments. Crops most commonly planted in decontamination systems in Colombia are plantain (Musa paradisiaca), papaya (Carica papaya), bore (Alocasia macrorrhiza), sugar cane (Saccharum officinarum) and nacedero tree (Trichanthera gigantea). They are com- monly used for forage production in Colombia. Under local conditions it produces about 10 tons of dry matter ha/year with 18 per cent of protein in the foliage dry matter. A good number of them grow very well in the sub-Saharan Africa, therefore are suggested for phytoremediation in that region.

There are two major forms of phytoextraction, namely assisted or natural phytoextraction. In induced or assisted phytoextraction, hyper-accumulators are cultivated for the purpose of remediation. It is associated with the use of chelators in soils to increase metal solubility or mobilization so that the plants can absorb them more easily. In natural phytoextraction, plants naturally take up the contaminants in soil unassisted. Many natural hyperaccumulators are metallophyte plants that can tolerate and incorporate high levels of toxic metals.

An advantage of phytoextraction is friendly moderate impact in the soil ecosystem. Most traditional methods commonly used for cleaning up heavy metal-contaminated soil disrupt soil structure and reduce soil productivity, but phytoextraction has the ability of cleaning up the soil without causing any kind of harm to soil quality and soil structural integrity. In addition to this, phytoextraction is cost-effective when compared with other soil remediation techniques, although it is frequently argued argued that significant effects are only achieved in the long term.

Phytotransformation describes chemical modification of environmental substances as a direct result of plant catabolic and anabolic activities. These activities lead to inactivation, degradation or immobilization. The degradation as caused by plants is referred to as phytodegradation, On the other hand, immobilization is known as phytostabilization which is a process of reducing the mobility of substances in the environment, for example, by limiting the leaching of substances from the soil.

Certain plants render organic pollutants, such as pesticides, explosives, solvents, industrial chemicals, and other xenobiotic substances non-toxic by their metabolism. Sometimes, microorganisms living in association with plant roots may metabolize these substances in soil or water. These complex and recalcitrant compounds cannot be broken down to basic molecules (water, carbon-dioxide, etc.) by plant molecules, and, hence, the term phytotransformation represents a change in chemical structure without complete breakdown of the compound. The term "Green Liver Model" is used to describe phytotransformation, as plants behave analogously to the human liver when dealing with these xenobiotic compounds or foreign compounds. After uptake of the xenobiotics, plant enzymes increase the polarity of the xenobiotics by adding functional groups such as hydroxyl groups (-OH).

This is known as Phase I metabolism, similar to the way that the human liver increases the polarity of drugs and foreign compounds. Whereas in the human liver enzymes such as Cytochrome P450s are responsible for the initial reactions, in plants enzymes such as nitroreductases carry out the same role.

In the second stage of phytotransformation, known as Phase II metabolism, plant biomolecules such as glucose and amino acids are added to the polarized xenobiotic to further increase the polarity (known as conjugation). This is again similar to the processes occurring in the human liver where glucuronidation (addition of glucose molecules by the UGT (e.g. UGT1A1) class of enzymes) and glutathione addition reactions occur on reactive centres of the xenobiotic.

Phase I and II reactions serve to increase the polarity and reduce the toxicity of the compounds, although many exceptions to the rule are seen. The increased polarity also allows for easy transport of the xenobiotic along aqueous channels.

In the final stage of phytotransformation (Phase III metabolism), a sequestration of the xenobiotic occurs within the plant. The xenobiotics polymerize in a lignin-like manner and develop a complex structure that is sequestered in the plant. This ensures that the xenobiotic is safely stored, and does not affect the functioning of the plant. However, preliminary studies have shown that these plants can be toxic to small animals (such as snails), and, hence, plants involved in phytotransformation may need to be maintained in a closed enclosure.Hence, the plants reduce toxicity (with exceptions) and sequester the xenobiotics in phytotransformation. Trinitrotoluene phytotransformation has been extensively researched and a transformation pathway has been proposed.

In the case of organic pollutants, such as pesticides, explosives, solvents, industrial chemicals, and other xenobiotic substances, certain plants, such as Cannas, render these substances nontoxic by their metabolism. In other cases, microorganisms living in association with plant roots may metabolize these substances in soil or water. These complex and recalcitrant compounds cannot be broken down to basic molecules (water, carbon-dioxide, etc.) by plant molecules, and, hence, the term phytotransformation represents a change in chemical structure without complete breakdown of the compound. The mechanism is likened to the Green Liver Model which is used to describe phytotransformation, as plants behave analogously to the human liver when dealing with these foreign compound/pollutant, After uptake of the xenobiotics, plant enzymes increase the polarity of the xenobiotics by adding functional groups such as hydroxyl groups (-OH).

This is known as Phase I metabolism, similar to the way that the human liver increases the polarity of drugs and foreign compounds. Whereas in the human liver enzymes such as Cytochrome P450s are responsible for the initial reactions, in plants enzymes such as nitroreductases carry out the same role. In the Phase II metabolism, plant biomolecules such as glucose and amino acids are added to the polarized foreign compound pollutants to further increase the polarity. This is known as conjugation and is again similar to the processes occurring in the human liver where glucuronidation and glutathione addition reactions occur on reactive centres of the xenobiotic.

Phase I and II reactions serve to increase the polarity and reduce the toxicity of the compounds, although many exceptions to the rule are seen. The increased polarity also allows for easy transport of the xenobiotic along aqueous channels. In the Phase III metabolism, the foreign pollutant compounds are a sequestered within the plant. The xenobiotics polymerize in a lignin-like manner and develop a complex structure that is sequestered in the plant where they are safely stored. However, such plants can be toxic to small animals like snails, and, hence, plants involved in phytotransformation may need to be maintained in a closed enclosure. Plants therefore reduce toxicity and sequester the xenobiotics through phytotransformation. Trinitrotoluene phytotransformation has been extensively researched and a transformation pathway has been proposed.

A significant number of organic chemicals and many inorganic ones are subject to enzymatic attack through the activities of living organisms. Efficacy of microbes in decontamination depends on some edaphic properties such as soil pH soil aeration, soil nutrient status, soil moisture, soil temperature, soil texture and type of heavy metal. According to Thapa et al. most of modern society's environmental pollutants are included among these chemicals, and the actions of enzymes on them are usually lumped under the term biodegradation. The productive use of biodegradative processes eliminate or detoxify pollutants that have found their way into the environment and threaten public health, usually as contaminants of soil, water, or sediments is bioremediation.

Some microbes can reduce activity of different types of heavy metals. Agricultural wastewater treatment can be effectively undertaken through biological processes involving the activity of microorganisms such as bacteria, algae, fungi, plants and animals. This they can do by their ability to convert active forms of toxic metals to inactive forms. However, choice of microbes depends on the availability of energy sources of the organisms in question. Other environmental conditions like temperatures, oxygen, moisture and the presence of hazardous contaminant contribute immensely in influencing efficacy of microbes in remediation programmes. The aerobic bacteria recognized for their degradative abilities are Pseudomonas, Alcaligenes, Sphingomonas. These microbes have often been reported to degrade pesticides and hydrocarbons, both alkanes and polyaromatic compounds. Many of these bacteria use the contaminant as the sole source of carbon and energy. The contact between the bacteria and contaminant is a precondition for degradation. Some bacteria are mobile and exhibit a chemotactic response, sensing the contaminant and moving toward it.

Soil fungi are very helpful in cleaning the pedosphere. The use of fungi in remediation is mycoremediation. Mycoremediation is a form of bioremediation in which fungi are used to decontaminate the area. The term mycoremediation refers specifically to the use of fungal mycelia in bioremediation. One of the primary roles of fungi in the ecosystem is decomposition, which is performed by the mycelium. The mycelium secretes extracellular enzymes and acids that break down lignin and cellulose, the two main building blocks of plant fiber. These are organic compounds composed of long chains of carbon and hydrogen, structurally similar to many organic pollutants. The key to mycoremediation is determining the right fungal species to target a specific pollutant. Certain strains have been reported to successfully degrade the nerve gases VX and sarin.

In one conducted experiment, a plot of soil contaminated with diesel oil was inoculated with mycelia of oyster mushrooms; traditional bioremediation techniques (bacteria) were used on control plots. After four weeks, more than 95% of many of the PAH (polycyclic aromatic hydrocarbons) had been reduced to non-toxic components in the mycelial-inoculated plots. It appears that the natural microbial community participates with the fungi to break down contaminants, eventually into carbon dioxide and water. Wood-degrading fungi are particularly effective in breaking down aromatic pollutants (toxic components of petroleum), as well as chlorinated compounds.

Rhizofiltration is the uptake of metals into plant roots. Mycofiltration is a similar process, using fungal mycelia to filter toxic waste and microorganisms from water in soil. Soils Arbuscular mycorrhizae (AM) are ubiquitous symbiotic associations between higher plants and soil fungi and their extra-radical mycelium form bridges between plant roots and soil, and mediate the transfer of various elements into plants. There is also a growing body of evidence that arbuscular mycorrhizal fungi can exert protective effects on host plants under conditions of soil metal contamination. Binding of metals in mycorrhizal structures and immobilization of metals in the mycorrhizosphere may contribute to the direct effects. Indirect effects may include the mycorrhizal contribution to balanced plant mineral nutrition, especially P nutrition, leading to increased plant growth and enhanced metal tolerance. It has been widely reported that ectomycorrhizal and ericoid mycorrhizal fungi can increase the tolerance of their host plants to heavy metals when the metals are present at toxic levels. The underlying mechanism is thought to be the binding capacity of fungal hyphae to metals in the roots or in the rhizosphere which immobilizes the metals in or near the roots and thus depresses their translocation to the shoots. Arbuscular mycorrhizal plants may exhibit much lower shoot concentrations of Zn and higher plant yields than non-mycorrhizal controls, indicating a protective effect of mycorrhizas on the host plants against potential Zn toxicity. It has been demonstrated that at high soil heavy metal concentrations, arbuscular mycorrhizal infection reduced the concentrations of Zn, Cd and Mn in plant leaves. Field investigations have indicated that mycorrhizal fungi can colonize plant roots extensively even in metal contaminated sites.

Phytodegradation is commonly applied as a phytoremediation measure. Phytodegradation (also rhizodegradation) is the breakdown of contaminants through the activity existing in the rhizosphere. Rhizobacteria are effective in nickel extraction. It is facilitated by the presence of proteins and enzymes produced by the plants or by soil organisms such as bacteria, yeast, and fungi. Rhizodegradation is a symbiotic relationship where the plants provide nutrients necessary for the microbes to thrive, while microbes provide a healthier soil environment.

Rhizofiltration is a water remediation technique that involves the uptake of contaminants by plant roots. Rhizofiltration is used to reduce contamination in natural wetlands and estuarine areas.

Phytodegradation or rhizodegradation is the breakdown of contaminants through the activity existing in the rhizosphere due to the presence of proteins and enzymes produced by the plants or by soil organisms such as bacteria, yeast, and fungi. Rhizodegradation is a symbiotic relationship where the plants provide nutrients necessary for the microbes to thrive, while microbes provide a healthier soil environment.

Soils that have been contaminated for a long time may undergo prolonged remediation and are less responsive to rhizodegradation than their freshly contaminated counterparts. There is therefore a need for enhancement of bioavailability as a key for successful biodegradation. Often times, selection and engineering of plants and microbial strains that modify solubility and transport of organic pollutants through exudation of biosurfactants become necessary and promising. In enhancing rhizodegradation, gene cloning of plants containing bacterial enzymes for the degradation of organic pollutants such as PCBs will be helpful in this regard. Other practices include the use of root-colonising bacteria like Pseudomonas fluorescens expressing degradative enzymes such as ortho-monooxygenase for toluene degradation. In many countries, soils and sediments polluted with crude oil hydrocarbons are of major environmental concern on various contaminated sites. Hydrocarbon-degrading microorganisms are ubiquitously distributed in soils and constitute less than 1% of the total microbial communities but may increase to 10% in the presence of crude oil. However, use of fertilizers in hydrocarbon-contaminated soils act as biostimulants in such conditions. Some microbes are able to use HC as a carbon and energy source preferentially in the absence of a readily available carbon source like labile natural organic matter. Read et al. observed increased phosphorus mobilisation due to exudation of biosurfactants by lupine (Lupinus angustifolius).

Rhizofiltration is a water remediation technique that involves the uptake of contaminants by plant roots. Rhizofiltration is used to reduce contamination in natural wetlands and estuary areas.

Bioremediation can be classified as ex situ and in situ bioremediation. The former techniques involve the excavation or removal of soil from ground. Important ex situ treatments are composting, biopiles land farming, and bioreactors. In situ is a simple technique in which contaminated soil is excavated and spread over a prepared bed and periodically tilled until pollutants are degraded. The goal is to stimulate indigenous biodegradative microorganisms and facilitate the aerobic degradation of contaminants. The practice is limited to the treatment of superficial 35 cm of soil. Since land farming has the potential to reduce monitoring and maintenance costs, as well as clean-up abilities, it has received much attention as a disposal alternative. In land farming, contaminated soils are combined with nonhazardous organic amendments such as manure or agricultural wastes. Organic materials in land farming supports the development of a rich microbial population and elevated temperature Composting is a process of piling contaminated soil organic substances such as manure or agricultural wastes. The added organic material supports the development of a rich microbial population and elevates temperature of the pile. Stimulation of microbial growth by added nutrients results in effective biodegradation in a relatively short period of time characteristic of composting. Sometimes, biopiles are used in bioremediation. A biopile is a hybrid of land farming and composting; and is used for treatment of surfaces contaminated with petroleum hydrocarbons. Biopiles are improved forms of land farming that tend to control physical losses of the contaminants through leaching and volatilization. Land farming is a method in which contaminated soil is spread over a prepared bed along with some fertilizers and occasionally rotated. It stimulates the activity of bacteria and enhances the degradation of oil. But, the use of biopiles provides a favourable environment for autochthonous aerobic and anaerobic microorganisms.

Composting is a process of piling contaminated soil organic substances such as manure or agricultural wastes. The added organic material supports the development of a rich microbial population and elevates temperature of the pile. Stimulation of microbial growth by added nutrients results in effective biodegradation in a relatively short period of time.

Most in situ bioremediation techniques are generally the most desirable options due to cheapness and fewer disturbances since they provide the services in place avoiding excavation and transport of contaminants. This could useful in pro-poor communities common in sub-Saharan Africa. However, in situ remediation is among other factors governed by depth of soils for its efficacy. In many soils effective oxygen is also a prerequisite. Examples of important in situ bioremediation include are biosparging, bioventing, in situ biodegradation, and bioaugmentation. The Deinococcus radiodurans is used for metal remediation in radioactively polluted environments.

Crude oil is a mixture of thousands of varying chemical compounds. Given that composition of each type of oil is unique, there are different ways to bioremediate them using microbes and flora. Bioremediation can occur naturally or can be encourage with addition of microbes and fertilizers.

The microbes present in the soil at early stage recognize the oil and its constituents by biosurfactants and bio emulsifiers. After this, they attach themselves and use the hydrocarbon present in the petroleum as a source of energy. However, low solubility and adsorption of high molecular weight hydrocarbons can pose as a limiting factor to their availability to microorganisms. But, addition of biosurfactants enhances the solubility and removal of these contaminants. Again, rates of oil biodegradations increases with addition of biosurfactants.

Volatility, volubility, and susceptibility to biodegradation differ distinctly among constituents of crude oil. Some compounds are easily degraded, some resist degradation and some are non-biodegradable. Yet, biodegradation of different petroleum compounds occurs simultaneously but at different rates because different species of microbes preferentially attack different compounds. This scenario leads to progressive and successive disappearance of constituents of crude oil over time.

Microbes produce enzymes in the presence of carbon sources, and these enzymes are responsible for the breakdown of hydrocarbon molecules. Many different enzymes and metabolic pathways are involved in the degradation of hydrocarbons contained in crude oil polluted soils. It implies that complete hydrocarbon degradation requires an appropriate enzyme, unavailability of which either prevents or minimizes its breakdown.

Bioremediation has various benefits of outstanding environmental and agricultural implications.

People perceive bioremediation as an acceptable strategy for the transformation of a wide variety of pollutants, often involving recycling.

Byproducts from bioremediation treatment are usually harmless products. Such residues include carbon dioxide, water, and cellular biomass, implying that most hazardous contaminants can be transformed to harmless products thereby eliminating the chance of future liability associated with treatment and disposal of contaminated material.

Processes involved in bioremediation can be conducted on-site without causing a major disruption of normal activities of the ecosystem. But, this, they need to transport quantities of waste off site and the potential threats to human health and the environment that can arise during transportation are eliminated.

Bioremediation is cheap when compared with other technologies that are used for clean-up of toxic waste. Some of the contaminants are sources of energy to the soil microbes thereby sustaining microbial biodiversity. Certain bacteria are mobile and exhibit a chemotactic response, sensing the contaminant and moving toward it.

Bioremediation was described as a strategy for integrated and sustainable development. More possibilities of recycling wastes within farming systems become available as wastes from one process become inputs for another.

Limitations of Bioremediation

- Bioremediation is limited to those compounds that can be degraded biologically. However, not all compounds are susceptible to rapid and complete degradation. Some substrates such as straw saw dust and maize cobs can be used to facilitate contact between soil microbes and toxicants.

- Biological processes are often highly specific. For instance, anaerobic bacteria used for bioremediation of polychlorinated biphenyls in river sediments, dechlorination of the solvent trichloroethylene. The white rot fungus Phanaerochaete chrysosporium have the ability to degrade an extremely diverse range of persistent or toxic environmental pollutants

- Research is needed to develop and engineer bioremediation technologies that are appropriate for sites with complex mixtures of contaminants that are not evenly dispersed in the environment.

- Bioremediation often takes longer than other treatment options, such as excavation and removal of soil or incineration.

- Regulatory uncertainty remains regarding acceptable performance criteria for bioremediation and there are no acceptable endpoints for bioremediation treatments.

- There are some concerns that the products of biodegradation may be more persistent or toxic than the parent compound.

Bioremediation of Heavy Metal Polluted Soils

Bioremediation is the use of organisms (microorganisms and/or plants) for the treatment of polluted soils. It is a widely accepted method of soil remediation because it is perceived to occur via natural processes. It is equally a cost effective method of soil remediation. Blaylock et al. reported 50% to 65% saving when bioremediation was used for the treatment of 1 acre of Pb polluted soil compared with the case when a conventional method (excavation and landfill) was used for the same purpose. Although bioremediation is a non-disruptive method of soil remediation, it is usually time consuming and its use for the treatment of heavy metal polluted soils is sometimes affected by the climatic and geological conditions of the site to be remediated.

Heavy metals cannot be degraded during bioremediation but can only be transformed from one organic complex or oxidation state to another. Due to a change in their oxidation state, heavy metals can be transformed to become either less toxic, easily volatilized, more water soluble (and thus can be removed through leaching), less water soluble (which allows them to precipitate and become easily removed from the environment) or less bioavailable.

Bioremediation of heavy metals can be achieved via the use of microorganisms, plants, or the combination of both organisms.

Using Microbes for Remediation of Heavy Metal Polluted Soils

Several microorganisms especially bacteria (Bacillus subtilis, Pseudomonas putida, and Enterobacter cloacae) have been successfully used for the reduction of Cr (VI) to the less toxic Cr (III). B. subtilis has also been reported to reduce nonmetallic elements. For instance, Garbisu et al. recorded that B. subtilis reduced the selenite to the less toxic elemental Se. Further, B. cereus and B. thuringiensis have been shown to increase extraction of Cd and Zn from Cd-rich soil and soil polluted with effluent from metal industry. It is assumed that the production of siderophore (Fe complexing molecules) by bacteria may have facilitated the extraction of these metals from the soil; this is because heavy metals have been reported to simulate the production of siderophore and this consequently affects their bioavailability. For instance, siderophore production by Azotobacter vinelandii was increased in the presence of Zn (II). Hence, heavy metals influence the activities of siderophore-producing bacteria which in turn increases mobility and extraction of these metals in soil.

Bioremediation can also occur indirectly via bioprecipitation by sulphate reducing bacteria (Desulfovibrio desulfuricans) which converts sulphate to hydrogen sulphate which subsequently reacts with heavy metals such as Cd and Zn to form insoluble forms of these metal sulphides.

Most of the above microbe assisted remediation is carried out ex situ. However, a very important in situ microbe assisted remediation is the microbial reduction of soluble mercuric ions Hg (II) to volatile metallic mercury and Hg (0) carried out by mercury resistant bacteria. The reduced Hg (0) can easily volatilize out of the environment and subsequently be diluted in the atmosphere.

Genetic engineering can be adopted in microbe assisted remediation of heavy metal polluted soils. For instance, Valls et al. reported that genetically engineered Ralstonia eutropha can be used to sequester metals (such as Cd) in polluted soils. This is made possible by the introduction of metallothionein (cysteine rich metal binding protein) from mouse on the cell surface on this organism. Although the sequestered metals remain in the soil, they are made less bioavailable and hence less harmful. The controversies surrounding genetically modified organisms and the fact that the heavy metal remains in the soil are major limitations to this approach to bioremediation.

Making the soil favourable for soil microbes is one strategy employed in bioremediation of polluted soils. This process known as biostimulation involves the addition of nutrients in the form of manure or other organic amendments which serve as C source for microorganisms present in the soil. The added nutrients increase the growth and activities of microorganisms involved in the remediation process and thus this increases the efficiency of bioremediation.

Although biostimulation is usually employed for the biodegradation of organic pollutants, it can equally be used for the remediation of heavy metal polluted soils. Since heavy metals cannot be biodegraded, biostimulation can indirectly enhance remediation of heavy metal polluted soil through alteration of soil pH. It is well known that the addition of organic materials reduces the pH of the soil; this subsequently increases the solubility and hence bioavailability of heavy metals which can then be easily extracted from the soil.

Biochar is one organic material that is currently being exploited for its potential in the management of heavy metal polluted soils. Namgay et al. recorded a reduction in the availability of heavy metals when the polluted soil was amended with biochar; this in turn reduced plant absorption of the metals. The ability of biochar to increase soil pH unlike most other organic amendments may have increased sorption of these metals, thus reducing their bioavailability for plant uptake. It is important to note that, since the characteristics of biochar vary widely depending on its method of production and the feedstock used in its production, the effect different biochar amendments will have on the availability of heavy metals in soil will also differ. Further, more research is needed in order to understand the effect of biochar on soil microorganisms and how the interaction between biochar and soil microbes influences remediation of heavy metal polluted soils because such studies are rare in literature.

Using Plants for Remediation of Heavy Metal Polluted Soils

Phytoremediation is an aspect of bioremediation that uses plants for the treatment of polluted soils. It is suitable when the pollutants cover a wide area and when they are within the root zone of the plant. Phytoremediation of heavy metal polluted soils can be achieved via different mechanisms. These mechanisms include phytoextraction, phyto-stabilization, and phytovolatilization.

Phytoextraction

This is the most common form of phytoremediation. It involves accumulation of heavy metals in the roots and shoots of phytoremediation plants. These plants are later harvested and incinerated. Plants used for phytoextraction usually possess the following characteristics: rapid growth rate, high biomass, extensive root system, and ability to tolerate high amounts of heavy metals. This ability to tolerate high concentration of heavy metals by these plants may lead to metal accumulation in the harvestable part; this may be problematic through contamination of the food chain.

There are two approaches to phytoextraction depending on the characteristics of the plants involved in the process. The first approach involves the use of natural hyperac-cumulators, that is, plants with very high metal-accumulating ability, while the second approach involves the use of high biomass plants whose ability to accumulate metals is induced by the use of chelates, that is, soil amendments with metal mobilizing capacity.

Hyperaccumulators accumulate 10 to 500 times more metals than ordinary plant ; hence they are very suitable for phytoremediation. An important characteristic which makes hyperaccumulation possible is the tolerance of these plants to increasing con-centrations of these metals (hypertolerance). This could be a result of exclusion of these metals from the plants or by compartmentalization of these metal ions; that is, the met-als are retained in the vacuolar compartments or cell walls and thus do not have access to cellular sites where vital functions such as respiration and cell division take place.

Generally, a plant can be called a hyperaccumulator if it meets the following criteria: (i) the concentration of metal in the shoot must be higher than 0.1% for Al, As, Co, Cr, Cu, Ni, and Se, higher than 0.01% for Cd, and higher than 1.0% for Zn; (ii) the ratio of shoot to root concentration must be consistently higher than 1; this indicates the capability to transport metals from roots to shoot and the existence of hypertolerance ability; (iii) the ratio of shoot to root concentration must be higher than 1; this indicates the de-gree of plant metal uptake. Reeves and Baker reported some examples of plants which have the ability to accumulate large amounts of heavy metals and hence can be used in remediation studies. Some of these plants include Haumaniastrum robertii (Co hyperaccumulator); Aeollanthus subacaulis (Cu hyperaccumulator); Maytenus bureaviana (Mn hyperaccumulator); Minuartia verna and Agrostis tenuis (Pb hyperaccumulators); Dichapetalum gelonioides, Thlaspi tatrense, and Thlaspi caerulescens (Zn hyperaccumulators); Psycotria vanhermanni and Streptanthus polygaloides (Ni hyperaccumulators); Lecythis ollaria (Se hyperaccumulator). Pteris vittata is an example of a hyperaccumulator that can be used for the remediation of soils polluted with As. Some plants have the ability to accumulate more than one metal. For instance, Yang et al. observed that the Zn hyperaccumulator, Sedum alfredii, can equally hyperaccumulate Cd.

The possibility of contaminating the food chain through the use of hyperaccumulators is a major limitation in phytoextraction. However, many species of the Brassicaceae family which are known to be hyperaccumulators of heavy metals contain high amounts of thiocyanates which make them unpalatable to animals; thus this reduces the availability of these metals in the food chain.

Most hyperaccumulators are generally slow growers with low plant biomass; this reduces the efficiency of the remediation process. Thus, in order to increase the efficiency of phytoextraction, plants with high growth rate as well as high biomass (e.g., maize, sorghum, and alfalfa) are sometimes used together with metal chelating substances for soil remediation exercise. It is important to note that some hyperaccumulators such as certain species within the Brassica genus (Brassica napus, Brassica juncea, and Brassica rapa) are fast growers with high biomass.

In most cases, plants absorb metals that are readily available in the soil solution. Although some metals are present in soluble forms for plant uptake, others occur as insoluble precipitate and are thus unavailable for plant uptake. Addition of chelating substances prevents precipitation and metal sorption via the formation of metal chelate complexes; this subsequently increases the bioavailability of these metals. Further, the addition of chelates to the soil can transport more metals into the soil solution through the dissolution of precipitated compounds and desorption of sorbed species. Certain chelates are also able to translocate heavy metal into the shoots of plants.

Marques et al. documented examples of synthetic chelates which have successfully been used to extract heavy metals from polluted soils. Some of these chelates include EDTA (ethylenediaminetetraacetic acid), EDDS (SS-ethylenediamine disuccinic acid), CDTA (trans-1,2-diaminocyclohexane-N,N,N′,N′-tetraacetic acid), EDDHA (ethylenediamine-di-o-hydroxyphenylacetic acid), DTPA (diethylenetriaminepentaacetic acid), and HEDTA (N-hydroxyethylenediaminetriacetic acid). EDTA is a synthetic chelate that is widely used not only because it is the least expensive compared with other synthetic chelates but also because it has a high ability to successfully improve plant metal uptake. Organic chelates such as citric acid and malic acid can also be used to improve phytoextraction of heavy metals from polluted soils.

One major disadvantage of using chelates in phytoextraction is the possible contamination of groundwater via leaching of these heavy metals. This is because of the increased availability of heavy metals in the soil solution when these chelates are used. In addition, when chelates (especially synthetic chelates) are used in high concentrations, they can become toxic to plants and soil microbes. In general, solubility/availability of heavy metals for plant uptake and suitability of a site for phytoextraction are additional factors that should be considered (in addition to suitability of plants) before using phytoextraction for soil remediation.

Phytostabilization

Phytostabilization involves using plants to immobilize metals, thus reducing their bioavailability via erosion and leaching. It is mostly used when phytoextraction is not desirable or even possible. Marques et al. argued that this form of phytoremediation is best applied when the soil is so heavily polluted so that using plants for metal extraction would take a long time to be achieved and thus would not be adequate. Jadia and Fulekar on the other hand showed that the growth of plants (used for phytostabilization) was adversely affected when the concentration of heavy metal in the soil was high.

Phytostabilization of heavy metals takes place as a result of precipitation, sorption, metal valence reduction, or complexation. The efficiency of phytostabilization depends on the plant and soil amendment used. Plants help in stabilizing the soil through their root systems; thus, they prevent erosion. Plant root systems equally prevent leaching via reduction of water percolation through the soil. In addition, plants prevent man's direct contact with pollutants and they equally provide surfaces for metal precipitation and sorption.

Based on the above factors, it is important that appropriate plants are selected for phytostabilization of heavy metals. Plants used for phytostabilization should have the following characteristics: Dense rooting system, ability to tolerate soil conditions, ease of establishment and maintenance under field conditions, rapid growth to provide adequate ground coverage, and longevity and ability to self-propagate.

Soil amendments used in phytostabilization help to inactivate heavy metals; thus, they prevent plant metal uptake and reduce biological activity. Organic materials are mostly used as soil amendments in phytostabilization. Marques et al. showed that Zn percolation through the soil reduced by 80% after application of manure or compost to polluted soils on which Solanum nigrum was grown.

Other amendments that can be used for phytostabilization include phosphates, lime, biosolids, and litter. The best soil amendments are those that are easy to handle, safe to workers who apply them, easy to produce, and inexpensive and most importantly are not toxic to plants. Most of the times, organic amendments are used because of their low cost and the other benefits they provide such as provision of nutrients for plant growth and improvement of soil physical properties.

In general, phytostabilization is very useful when rapid immobilization of heavy metals is needed to prevent groundwater pollution. However, because the pollutants remain in the soil, constant monitoring of the environment is required and this may become a problem.

Phytovolatilization

In this form of phytoremediation, plants are used to take up pollutants from the soil; these pollutants are transformed into volatile forms and are subsequently transpired into the atmosphere. Phytovolatilization is mostly used for the remediation of soils polluted with Hg. The toxic form of Hg (mercuric ion) is transformed into the less toxic form (elemental Hg). The problem with this process is that the new product formed, that is, elemental Hg, may be redeposited into lakes and rivers after being recycled by precipitation; this in turn repeats the process of methyl-Hg production by anaerobic bacteria.

Raskin and Ensley reported the absence of plant species with Hg hyperaccumulating properties. Therefore, genetic engineered plants are mostly used in phytovolatilization. Examples of transgenic plants which have been used for phytovolatilization of Hg polluted soils are Nicotiana tabacum, Arabidopsis thaliana, and Liriodendron tulipifera. These plants are usually genetically modified to include gene for mercuric reductase, that is, merA. Organomercurial lyase (merB) is another bacterial gene used for the detoxification of methyl-Hg. Both merA and merB can be inserted into plants used to detoxify methyl-Hg to elemental Hg. Use of plants modified with merA and merB is not acceptable from a regulatory perspective. However, plants altered with merB are more acceptable because the gene prevents the introduction of methyl-Hg into the food chain.

Phytovolatilization can also be employed for the remediation of soils polluted with Se. This involves the assimilation of inorganic Se into organic selenoamino acids (selenocysteine and selenomethionine). Selenomethionine is further biomethylated to dimethylselenide which is lost in the atmosphere via volatilization. Plants which have successfully been used for phytovolatilization of soils polluted with Se are Brassica juncea and Brassica napus.

Combining Plants and Microbes for the Remediation of Heavy Metal Polluted Soils

The combined use of both microorganisms and plants for the remediation of polluted soils results in a faster and more efficient clean-up of the polluted site. Mycorrhizal fungi have been used in several remediation studies involving heavy metals and the results obtained show that mycorrhizae employ different mechanisms for the remediation of heavy metal polluted soils. For instance, while some studies have shown enhanced phytoextraction through the accumulation of heavy metals in plants, others reported enhanced phytostabilization through metal immobilization and a reduced metal concentration in plants.

In general, the benefits derived from mycorrhizal associations—which range from increased nutrient and water acquisition to the provision of a stable soil for plant growth and increase in plant resistance to diseases — are believed to aid the survival of plants growing in polluted soils and thus help in the vegetation/revegetation of remediated soils. It is important to note that mycorrhiza does not always assist in the remediation of heavy metal polluted soils and this may be attributed to the species of mycorrhizal fungi and the concentration of heavy metals. Studies have also shown that activities of mycorrhizal fungi may be inhibited by heavy metals. In addition, Weissenhorn and Leyval reported that certain species of mycorrhizal fungi (arbuscular mycorrhizal fungi) can be more sensitive to pollutants compared to plants.

Other microorganisms apart from mycorrhizal fungi have also been used in conjunction with plants for the remediation of heavy metal polluted soils. Most of these microbes are the plant growth-promoting rhizobacteria (PGPR) that are usually found in the rhizosphere. These PGPR stimulate plant growth via several mechanisms such as production of phytohormones and supply of nutrients, production of siderophores and other chelating agents, specific enzyme activity and N fixation, and reduction in ethylene production which encourages root growth.

In general, PGPR have been used in phytoremediation studies to reduce plant stress associated with heavy metal polluted soils. Enhanced accumulation of heavy metals such as Cd and Ni by hyperaccumulators (Brassica juncea and Brassica napus) has been observed when the plants were inoculated with Bacillus sp. On the other hand, Madhaiyan et al. reported increased plant growth due to a reduction in the accumulation of Cd and Ni in the shoot and root tissues of tomato plant when it was inoculated with Methylobacterium oryzae and Burkholderia spp. Thus, this indicates that the mechanisms employed by PGPR in the phytoremediation of heavy metal polluted soils may be dependent on the species of PGRP and plant involved in the process. Although studies involving both the use of mycorrhizal fungi and PGPR are uncommon, Vivas et al. reported that PGPR (Brevibacillus sp.) increased mycorrhizal efficiency which in turn decreased metal accumulation and increased the growth of white clover growing on a heavy metal (Zn) polluted soil.

Control of Soil Pollution

Soils are considered as purification of the nature. Moreover supplying food, soils have also purification property. This soil property is caused by their physical properties (water permeability operation from pores), their chemical properties (surface absorption and evaporation) and their biological properties (decomposition and corruption of organic matters).

Controling Oil Pollutions in Soil

Oil materials and their derivatives may cause soil pollution as a result of transport or storage. If more oil materials are penetrated into the more depth of soil, removing its pollution will be more difficult. Some bacteria and microorganisms in soil can cause decomposition of oil materials.

Here under are regarded as the ways to control effects of oil pollution:

- Preventing oil from spreading widely,
- Improve the soil ventilation through plowing and mixing,
- Increasing food nutrients to the soil like nitrogen and phosphor,
- Combining soil with microorganisms which decompose oil materials.

Controlling Pollutions caused by Waste in Soil

Waste is one of the most important sources of soil contamination. Waste can penetrates into the ground and pollute the water resources as well.

Methods of waste disposal include: Dumping, incinerating and recycling.

In dumping method, areas are created as "Land field" and garbage or waste is dumped there. In this method, waste is dumped below the ground level, aimed at not to be observed from ground surface but the said method creates subsequent problems. These problems include: pollution of water resources, producing bad odor and poisonous methane gas which provides fire danger, accumulation of harmful insects and organisms.

To control soil pollution caused by the waste, the following techniques are recommended:

- Application of effective technology for dumping waste like compressing and covering of openings and holes,
- Dumping waste higher than the highest underground water levels,
- Creating impenetrable layers in building of land fields,
- Creating drainage system for the collection of leachates,
- Using the gases produced in land fields.

In incineration method, all wastes are collected at a place away from residing place and then, they are put on fire.

Incineration method is one of the worst methods of waste disposal, because, incineration will produce very poisonous gases which will pollute the air and incur irreparable loss to the environment as well. After incineration, ash of waste is remained and will create visual pollution.

Recycling is the best method of waste disposal. With storing some waste and re-using them, human can greatly contribute reduction of waste. In this method, not only creation of more waste is prevented, but also, more cost will also be saved remarkably.

Controling Pollution caused by Industrial Activities in Soil

This method includes all the pollutants which are entered the soil by the factories. These wastes include as follows: Wastes produced by steel industries and power plants, wastes of chemical industries, wastes of steel mfg. industries, wastes of metalworking industries, wastes of oil industries (extraction and refining), wastes of wood, cellulose and paper mfg. industries, wastes of leather production industries as well as waste of food industries.

Accumulation of heavy metals in soil is the major discussion of the industrial pollutions. These metals include lead, cadmium, silver and mercury which their harmful effects have been proven on the living organisms and have repeatedly caused environmental disasters.

Some of these effects are as follows:

- Disturbance of biological activities of soil,

- Toxic effects on plants,

- Harmful effects on human being as a result of entrance of materials to the food chain.

There are three main methods for soil decontamination from industrial wastes as follows:

- Soil can be excavated up to the specified depth and the excavated soil can be taken away from the region and then, it can be restored.

- The soil can be restored at the same area.

- Keeping soil in the area is the other method. Under such circumstances, auxiliaries are added to the soil to prevent spread of infection to the plants, animals and human.

Usually, a large plastic is drawn on the soil to prevent spread of soil pollution, to prevent water from penetrating into the soil and to prevent spread of pollution to the other regions.

The Soil Restoration methods includes as Follows

Using water to remove pollutants from the soil, using chemical and aerial solvents, eliminating pollutants with incineration, helping natural organisms for breaking down atoms of pollutants, adding materials to the soil for protecting it and preventing spread of pollution to the other regions.

Role of Plants in Controlling and Decreasing Soil Pollution

Pollution caused by the exhaust of cars and homes and departments' heating devices and appliances will incur irreparable damages to the health of human beings, animals, water, soil and air.

Increase of diseases in human beings, eradication of many plants and endangerment of many generations of animals are a solid evidence of the said claim.

For this purpose, biological scientists and environmentalists forced to think of natural ways (biologic) for fighting with pollutions. The use of plants is the simplest biological way, because, using plants is a non-technical and cost effective method in terms of technology.

Soil Pollution Control Technique caused by Lead Existing in it

Fungi are used to fight lead existing in soil, because, satisfactory and good reports have been received both in coexistence among plants and fungi like Arbascular – Mycorriza Fungi (AMF) in absorption of the lead. Creating colony of this fungus on the root of plants will cause increase of root level for absorption of the lead which results in more absorption of the lead element to the host plant. These reports indicate that these fungi help plants survive and tolerate pollution better. The researchers consider this tolerance as a result of protection of their roots by the fungi in soil.

The protective effect of fungi for the plants is as follows: Mycoriza colony is increased on root and all surfaces of the plant with the aim of absorbing heavy metals like lead. Even if the plant is able to accumulate a lot of these elements, the fungus plays an intermediary role for absorption of heavy metals again.

In lead mines, a type of fungus named Spurs Plows Niger can also be sued. As compared with the other fungi, this fungus can absorb more lead from the environment without experiencing any damage in this respect. So, these types of fungi can help improve soil infected to the heavy metals like lead.

Phytoremediation Technique to Control Soil Pollution

Phytoremediation is a cost-effective, environmental and scientific technique which is suitable for developing countries and is considered as a valuable business. Unfortunately, despite this potential, the technique, as a technology, has not yet commercially used in some countries like our country. Through the use of Green Plants Engineering like herbaceous and woody species, phytoremediation is used for removing pollutants from water and soil or decreasing risks of environmental pollutants like heavy metals, rare elements, organic compounds and radioactive materials.

Heavy metals are the most important mineral pollutants and soil microorganisms are able to decompose organic pollutants. But for microbial decomposition of metals, there is a need for organic or metal changes, in which, plants are presently used for this part.

Although way of using the plants, which are polluted in this form, is the major concern of the experts, strategy of generating energy opened another chapter for the scientists as one of the most essential aspects of today's life.

From a global perspective, soil crust is considered as the third major component of the human environment after water and air. In addition to the base of terrestrial living organism especially human communities, soil is considered as a unique environment for living of different life especially plants. Unlike the weather, soil pollution cannot be measured easily in chemical compound terms. In other words, a clean or pure soil is indefinable. Then, we have to study potential issues of the soil pollution within the framework of anticipated probable damages and hazards in performance of soil. With the development of manmade projects and contamination of soils by the heavy metals, the structure of soil will be dangerous and poisonous for the growth and development of plants and will entangle biodiversity of the soil as well.

In phytoremediation method, plants are classified based on absorption mechanism and soli pollution to the heavy metals is reduced through chemical, physical and biological methods. Based on the researches made by the Bureau to Study Water and Soil Pollution of the Department of the Environment (DoE), removal of soil pollution is usually carried out by two methods: 1. inside the site and 2. outside the site. In outside-of-the-site method, the contaminated soil is transferred to the other place and after removal of its pollution, it is returned to the first place. In another method i.e. inside-the-site, there is no need to transfer or move the soil, rather, when pollutants are turned into organic, capability of their biological uptake is reduced. To reduce pollution of mineral pollutants in soil, the following methods of organic, complex and soil increase by the lime can be used.

But most of these methods are expensive and will destruct the environment. In technology of using plants as phytoremediation, green plants and their relationship with soil microorganisms is used to reduce soil and groundwater pollution. This technology can be used for removing each of two pollutants of soil i.e. organic and inorganic.

The studies show that application of physicochemical techniques will cause eradication of soil useful microorganisms like microriza nitrogen stabilizers and consequently, it will weaken biological activities of the soil which will cost dearly in comparison with the phytoremediation technique.

Soil and water plants are used in rizo-filtration method that pollutants of contaminated water resources are deposited or condensed with little density in their roots. This method is applied especially for industrial wastewater treatment plants, agricultural runoff and/or wastewater of acid mines and is suitable for the metals like lead, cadmium, copper, nickel, zinc and chrome. The plants like Indian mustard, sunflower, tobacco, rye and corn enjoy this capability. These plants enjoy high capability of absorbing lead from the sewage system, based on which, sunflower has the highest capability of absorbing lead from the sewage than other plants.

In another method, restricting mobility and availability of pollutants is carried out in soil through using power of the root. This method is usually used in soil, sediment and sludge to reduce the pollution and is carried out through absorption, deposition, complex and/or reduction of the capacity.

In plant evaporation method, plants absorb pollutants from the soil and then convert them into the steam and transfer with the transpiration and atmosphere operation. This method is used in growing trees to absorb organic and inorganic pollutants. In another method which is known as "Reduced Plan", the plant helps removal of pollution from soil and underground waters with its metabolism through transferring, decomposition, stabilization and sublimation of the pollutant compounds. In this method, organic compounds are broken into simpler molecules and can be entered inside the plant tissue. The studies have shown that plants enjoy the enzymes that can decompose waste of chlorinated solvents like tri-chlorine, ethylene and other insecticides.

Heavy Metals and their Availability in Soil

Heavy metals are the elements with atomic weight, ranging from 63.54 to 200.59, and specific weight more than 4. Some heavy metals are required little amount of living organisms although excessive increase of the same essential heavy metals can be harmful for the organisms. The unnecessary heavy metals include arsenic, antimony, cadmium, chrome, mercury and lead. These metals are very important with relation to the pollution of soil and surface waters and are taken into consideration in phytoremediation science.

Reaction of Plants to Heavy Metals

Plants have three basic strategies for the growth in soils contaminated to the heavy metals. 1. The species which prevent entrance of metals to their aerial parts or hold concentration of metals low in the soil. 2. The species which accumulate metals in their aerial organs and return them to the soil. 3. The plants which can condensate metals in their aerial organs and also the plants which absorb high condensation of pollutants and condense in their root, stem or leaf.

Application of Genetic Engineering to Improve Phytoremediation

In this technique, tolerance of species can be increased to the environmental pollutant metals through the genetic diversity existing inside each species and stimulation of genetic characteristics of the species. The studies have shown that production of plants with high potential of phytoremediation and biomass production is effective to improve phytoremediation method. The inoculation of the effective genes in accumulation of the metals to the plants which are taller than the natural plants will cause increase of final biomass production.

Phytoremediation methods have some limitations, for example, in a type of this method, phytoremediation method is effective at the range of three-feet from soil surface and maximum 60-feet from the underground waters. This method is applied vastly for the places which pollution density in them is between low to average level and is severely dependent on the soil acidity.

The results of research activities made by scientists indicate that soil acidity increases availability of metals to a great extent. Of course, acidic soil may have negative effects. For example, increased solubility of some toxic metals and washing them to the underground waters will cause outbreak of environmental hazards which should be controlled under specific conditions.

Consuming Phytoremediation Products

The way of consuming contaminated plants is one of the barriers to commercial implementation of phytoremediation. After harvesting, soil pollution is reduced by the plant but a great amount of dangerous biomass is produced. The studies show that compost production and condensation are the two methods which were proposed by many researchers for the biomass management of the contaminated plants. Thermo-chemical change is the best method for consuming biomasses produced by phytoremediation. As an energy source, biomass has commercial usage in this method. This biomass includes carbon, hydrogen and oxygen which are known as oxygenated hydrocarbons. Hemicelluloses, cellulose, minerals and ash is the main component of each lignin biomass which enjoy a high amount of humidity, volatile materials and bulk density but have low thermal value. The percent of these components varies from species to species, in which, management of this volume of waste is a difficult task and needs volume reduction.

Energy Production

Incinerating and producing gas is the important methods for generating thermal and electrical energy which can be extracted from the infected plants. Recycling this energy from biomass, by incinerating or producing gas, can have economic value, because, it cannot be consumed as fodder or fertilizer. Incinerating is a simple method which should be put under strict controlled situations. In this method, biomass is decreased between 2 to 5 percent and ash can be consumed appropriately. The studies have shown that incinerating dangerous wastes, containing metals, is not correct and logical in an open-air space, because, the gases released to the environment may be harmful. Under such circumstances, volume is only reduced and the produced heat is wasted.

The studies made by the Bureau to Study Water and Soil Pollution of the Department of the Environment (DoE) show that gas production is one of the cases for controlling biomass which is produced through a series of chemical changes of the clean combustion gases with high thermal efficiency. So, this gas combination is called "Pyro Gas" which can be incinerated for producing thermal and electrical energy.

Gas production is carried out in a gas exchanger within complicated stages of drying, heating, thermal decomposition and chemical reactions of combustion which is occurred simultaneously. The researchers have reported that pyrolysis is a new method for the management of urban wastes and it is possible to use it for biomass management of the contaminated plants. Pyrolysis decomposes materials under anaerobic situations and has no air emissions.

Thus, heavy metals remain in the coke which can be used in the melting furnace. Although high cost of establishment and operating procedures is a restricting factor, if used only for the plant, it can be suitable for the pollutant plants and urban wastes. The researchers have conducted research activities on plant species with high biomass and have shown that this method can have both positive results and effective environmental benefits.

Removal of Soil Oil Pollutions through using Phytoremediation Technique

Oil pollution is an inevitable consequence of rapid population growth and industrialization process. In the same direction, soil pollution by the oil hydrocarbon materials can be observed vastly around exploration and refining installations typically via transfer routes of these materials across southern oilfield provinces of the country.

In addition to the direct emission of these pollutants, the dusts from burning gases with oil have added toxic and harmful materials to the soils of the region for many years. Moreover having vast impact on ecosystem of the region, these pollutants are penetrated into the food chain and also human communities gradually and consequently, threaten health of human beings.

Presently, dire need is felt to prevent spread of these pollutions to the region and also cleaning up the contaminated regions.

For this purpose, different methods can be used. Phytoremediatin is one of these methods which take advantage of plants and microorganisms along with them for cleaning up the contaminated environments.

In fact, phytoemediation will provide suitable conditions for the growth and deployment of the plant as well as increase of natural cleanup activities through the application of human interventions such as agricultural technology (plowing and fertilizing). Cost-effectiveness, inexpensiveness and its simplicity is the most important advantage of this method than any other methods. In this method, selection of suitable plant is of paramount importance which depends on the climatic condition of the region, type and degree of soil pollution.

Future of Phytoremediation Technique in Controlling Soil Pollution

Although this science is developing very rapidly, the studies show that commercial phytoremediation should be able to compete with the other technologies in terms of time. More phytoremediation tests have been carried out in hydroponics environment in laboratory scale and heavy metals have been given to them, while soil environment is completely different. In real soil, there are many metals in insoluble forms and their availability is low. The said issue is the greatest problem. Many plants have not yet been known that should be identified and more should be known about their physiology.

Although 10 years have passed from the initial application of the phytoremediation technology in the world, this science has been developed very rapidly and today, phytoremediation has vast application on organic, inorganic and radioactive materials. This process is sustainable, affordable and is suitable for the developing countries. Generally, this method is cost effective as well. The studies show that efficiency of this method is increased with the application of fast-grown plants with high biomass and high absorption power of the heavy metals. In most contaminated places, appropriate species have been identified for the removal of pollution. Two methods of composting and condensing can be regarded as preliminary stages for reducing production volume of these plants but it should be considered that leachate of condensation should be collected completely. The researchers believe that incinerating consumes the least possible time among the methods which reduce the pollutants' biomass and is more suitable in comparison with the direct incineration in environmental terms.

However, it is observed that today world can devise improvements with inspiration of the nature and its virgin non-diminishable system for what the human being has destroyed with his hand which undoubtedly is not easier than preventing pollution of resources especially soil resources.

References

- Soil-remediation-types-techniques: spokaneenvironmental.com, Retrieved 16 June, 2019

- Remediation-Technologies-for-Contaminated-Sites- 278658230: researchgate.net, Retrieved 18 July, 2019

- Selected-bioremediation-techniques-in-polluted-tropical-soils, environmental-risk-assessment of-soil-contamination: intechopen.com, Retrieved 16 April, 2019

- Soil-Pollution-Control-Management-Techniques-and-Methods- 320395375: researchgate.net, Retrieved 08 August, 2019

5

Impact of Soil Pollution

Soil pollution adversely affects the wildlife, humans and the environment. It causes diseases like leukaemia, nervous system damage, etc. in humans, destruction of wildlife and its habitat, impact on soil fertility and agriculture, etc. All these impacts of soil pollution have been carefully analysed in this chapter.

Soil pollution impacts the environment and human health in three ways. First, contaminants attach themselves to dry soil particles and are blown away by the wind. These particles can then be inhaled by humans, who ingest the harmless soil particle as well as the more dangerous contaminant that is attached.

Second, water dissolves some of the contaminants in the soil, either during rainy weather or through groundwater action. Humans and animals then drink this contaminated water and ingest the contaminants.

Third, plants growing on polluted soil take up contaminants from the soil and store them in plant tissue. Humans and animals eat the plants and ingest the dangerous contaminants. For the second and third cases, humans that eat contaminated animals are also exposed to even more concentrated pollutants. Understanding how soil is polluted and what the consequences are is important for starting to work toward solutions.

Workings of Soil Contamination

Soil contamination is often present in areas that have seen a high level of industrial activity. Oil spills or chemical spills can contribute to the level of pollution. On farmland, farmers may have used toxic pesticides or fertilizers that contain harmful chemicals. Landfills can leach all kinds of chemicals into the surrounding soils and fires often add layers of toxic ash to exposed soils.

When contaminated soil particles drift through the air or flow along with water, humans can inhale or drink the pollutants attached to the particles. Traces of polluted soil on food may be eaten as well. Any of the pollutants in soils can enter the human body in this way.

Spreading the contaminants through drinking water or food plants and animals is more limited. Soil particles are generally filtered out of drinking water, and for contaminants to pass through the filters or to enter food plants and animals, they have to be soluble in water. Many industrial chemicals are only slightly water soluble and can't be spread easily in this way.

Soil Pollution Effects

The direct effects of soil pollution negatively influence the whole social and natural environment. Plants that grow on polluted soil may have lower yields because the hazardous chemicals in the soil interfere with their growth. Animals that eat polluted soil particles or contaminated plants may also grow more slowly or succumb to disease. Human health is impacted as well.

People who ingest polluted soil particles or who eat contaminated plants and animals may be poisoned by the chemical that enters their body. For example, lead is a common contaminant for areas with intensive industrial activity, and people may exhibit signs of lead poisoning. Other chemicals cause rashes and allergies, and they may weaken the immune system. All these effects depend on the type of chemical introduced into the body, how toxic it is and what its concentration is.

Soil Pollution Consequences

As soils around the world become more polluted, the consequences for society can be severe. Food shortages are likely due to the increasingly poor yields of farms. Some chemicals may reduce the overall health of large parts of the human population, leading to increased mortality and higher medical costs. When allergies become more common, the quality of life for many people is reduced, and weak immune systems may make people more susceptible to outbreaks of infectious diseases.

Healthy, productive and pure soil is essential to a healthy ecosystem and an environmentally sound social system. Once the pervasiveness, the effects and the consequences of soil pollution are widely known, people will have to work toward finding solutions and restoring pure soils wherever possible.

Diseases Caused by Soil Pollution

Soil pollution occurs when there is a build-up of persistent toxic compounds, salts, radioactive materials, chemicals or disease-causing agents in the soil which affect human, animal and plant health. Soil pollution is mainly a result of human activity, such as the application of pesticides like Atrazine, which is a popular weedkiller, and the generation of unwanted industrial waste like arsenic. Soil pollution changes the composition of the soil and creates a pathogenic soil environment, leading to the spread of diseases.

Cancer

Pesticides, benzene, chromium and weed killers are carcinogens which have been established to lead to all kinds of cancer. Long-term benzene exposure is responsible for irregular menstrual cycles in women, leukemia and anemia. A high level of exposure to benzene is fatal. Benzene is a liquid chemical found in crude oil, gasoline and cigarette smoke. It is used in chemical synthesis and interferes with cellular function by decreasing the production of red blood cells, white blood cells and antibodies, thereby compromising the body's immunity.

Kidney and Liver Disease

People develop kidney damage when they are exposed to soil which has been contaminated with lead. Soil pollutants like mercury and cyclodienes also greatly increase the possibility of developing irreversible kidney damage. Cyclodienes and PCBs cause toxicity in the liver, as well. This situation is worse for impoverished people who are forced by strained circumstances to live near dump sites, industrial factories and landfills, where they are exposed to soil pollution on a daily basis. They develop impaired immune systems, kidney damage and liver damage, in addition to neurological damage and lung problems.

Brain and Nerve Damage

Children can be exposed to the harmful effects of soil pollution in places like playgrounds and parks, where lead-contaminated soil has been proven to cause brain and neuromuscular development problems.

Malaria

Contaminated water or raw sewage may mix with soil in areas where the rainfall is usually heavy, such as in the tropics. The protozoa that cause malaria and the mosquitoes that act as carriers thrive in such conditions; the resulting increased propagation of both the protozoa and the mosquitoes leads to frequent outbreaks of malaria.

Cholera and Dysentry

Soil pollution is closely linked to water pollution, because when the soil is contaminated, it leaches into surface and ground water, leading to the contamination of drinking water and an outbreak of water-borne diseases like cholera and dysentery.

Impact of Soil Pollution on Soil Fertility

Soil pollution has adverse effects on the well-being of humans, plants and animals. Due to flood water, over-utilization of chemical fertilizers, pesticides and bacterial killers and excessive use of land due to multi-cropping system, so many distortions occur in the soil that it tends to lose its essential structure and element to maintain its fertility. The side effects of soil pollution include severe depletion of certain types of calcium, methane, nitrogen, sulphur, iron, copper, nitrogen, potassium and phosphorus in the soil; destruction of some useful plants and organisms which give moisture to the soil; toxic grains, vegetables and fruit that pose health problems to humans and alkalinity in the soil.

Soil pollution can reduce the yields of crops and plants. It causes the loss of soil and natural nutrients, leading to decrease in crop production. The 'chemical farming' adopted for green revolution or advanced yield is rather wasting it permanently instead of increasing soil fertility.

The physical and chemical properties of the soil are affected by soil pollution. Generally solid waste is buried under the soil. This leads to the loss of non-renewable metals such as copper, zinc, lead etc, and adverse impact on the production capacity of the soil. People sometimes irrigate farms with sewage water. This reduces the number of holes present in the soil day by day. Later, a situation comes that the natural sewage treatment system of the land is completely destroyed. When the land reaches such a situation, it is called a sick land.

Soil pollution has negative effects on plants and flora as well as the organisms that depend upon them.

Micro-organisms

Acidic soils created by deposit of acidic compounds such as sulfur dioxide produce acidic environment that is not tolerated by micro-organisms, which improve the soil structure by breaking down organic material and aiding in water flow.

Photosynthesis

Soils polluted by acid rain have an impact on plants by disrupting the soil chemistry and reducing plants' ability to take up nutrients and undergo photosynthesis.

Aluminum

While aluminum occurs naturally in the environment, soil pollution can mobilize inorganic forms, which are highly toxic to plants and can potentially leach into ground water, compounding their effects.

Algal Blooms

Contaminated soils with high levels of nitrogen and phosphorus can leach into waterways, causing algal blooms, resulting in the death of aquatic plants due to depleted dissolved oxygen.

pH

Acidic deposition into the soil can hamper its ability to buffer changes in the soil pH, causing plants to die off due to inhospitable conditions.

After-effects of Soil Pollution

Soil pollution is not only the problem of any one country but it is a global problem. It causes harmful effect on the soil and the environment at large. Contamination of soil will decrease the agricultural output of a land. Major soil pollution after effects are:

Inferior Crop Quality

It can decrease the quality of the crop. Regular use of chemical fertilizers, inorganic fertilizers, pesticides will decrease the fertility of the soil at a rapid rate and alter the structure of the soil. This will lead to decrease in soil quality and poor quality of crops. Over the time the soil will become less productive due to the accumulation of toxic chemicals in large quantity.

Water Sources Contamination

The surface run-off after raining will carry the polluted soil and enter into different water resource. Thus, it can cause underground water contamination thereby causing water pollution. This water after contamination is not fit for human as well as animal use due to the presence of toxic chemicals.

Negative Impact on Ecosystem and Biodiversity

Soil pollution can cause an imbalance of the ecosystem of the soil. The soil is an important habitat and is the house of different type of microorganisms, animals, reptiles, mammals, birds, and insects. Thus, soil pollution can negatively impact the lives of the living organisms and can result in the gradual death of many organisms. It can cause health threats to animals grazing in the contaminated soil or microorganisms residing in the soil.

Therefore, human activities are responsible for the majority of the soil pollution. We as humans buy things that are harmful and not necessary, use agricultural chemicals (fertilizers, pesticides, herbicides, etc.), drop waste here and there. Without being aware we harm our own environment.

Therefore, it is very important to educate people around you the importance of environment if they are not aware. Prevention of soil erosion will help to cease soil pollution. Thus, it is our small steps and activities that can help us to achieve a healthier planet for us. Therefore, it is essential for industries, individuals and businesses to understand the importance of soil and prevent soil pollution and stop the devastation caused to plant and animal life.

Soil Acidification

Soil acidification is a process where the soil pH decreases over time. This process is accelerated by agricultural production and can affect both the surface soil and subsoil.

Queensland has more than 500,000 hectares of agricultural and pastoral land that is acidified or is at risk of acidification. The higher-rainfall coastal areas used for intensive agriculture are most at risk.

Contributing Factors

Some contributing factors to soil acidification include:

- The application of high levels of ammonium-based nitrogen fertilisers to naturally acidic soils.
- Leaching of nitrate nitrogen, originally applied as ammonium-based fertilisers.
- Harvesting plant materials (plant material is alkaline so when it is removed the soil is more acidic than if the plant material had been returned to the soil).

Effects

Excessively acidic soils may lead to a dramatic decline in crop and pasture production because the pH of the soil changes the availability of soil nutrients.

Acidic soils may have some or all of the following problems:

- Helpful soil micro-organisms may be prevented from recycling nutrients (e.g. nitrogen supply may be reduced).
- Phosphorus in the soil may become less available to plants.
- Deficiencies of calcium, magnesium and molybdenum may occur.
- The ability of plants to use subsoil moisture may be limited.
- Aluminium, which is toxic to plants and micro-organisms, may be released from the soil.
- Levels of manganese may reach toxic levels.
- Uptake by crops and pastures of the heavy metal contaminant, cadmium, may increase.

Control Measures

It is most important that soil acidity be treated at an early stage. If acidity spreads into the subsoil, serious yield reduction may occur. Subsoil acidity is difficult and costly to control.

There are a number of ways to minimise the soil acidification process, including:

- The use of less acidifying farming practices—considered when soils show signs of acidification.

- Applications of agricultural lime—applied to counter the acidification caused by cropping systems.

Soil Salinization

Soil salinization is a serious and difficult to reverse form of soil degradation. Salinization occurs when dissolved salts in water tables rise to the soil surface and accumulate as water evaporates. Often rise in a water table is due to the replacement of deep-rooted vegetation, such as trees, with shallower rooted vegetation, such as grasses. Application of irrigation water or heavy rainfall can also cause water tables to rise. Topsoil salts can greatly reduce agricultural productivity, erode infrastructure, and impose long-term limitations on land productivity. Soils containing high levels of salts are much more likely to experience this regime shift.

Soil salinity is an enormous problem for agriculture under irrigation. In the hot and dry regions of the world the soils are frequently saline with low agricultural potential. In these areas most crops are grown under irrigation, and to exacerbate the problem, inadequate irrigation management leads to secondary salinization that affects 20% of irrigated land worldwide. Irrigated agriculture is a major human activity, which often leads to secondary salinization of land and water resources in arid and semi-arid conditions. Salts in the soil occur as ions (electrically charged forms of atoms or compounds). Ions are released from weathering minerals in the soil. They may also be applied through irrigation water or as fertilizers, or sometimes migrate upward in the soil from shallow groundwater. When precipitation is insufficient to leach ions from the soil profile, salts accumulate in the soil resulting soil salinity. All soils contain some water-soluble salts. Plants absorb essential nutrients in the form of soluble salts, but excessive accumulation strongly suppresses the plant growth. During the last century, physical, chemical and/or biological land degradation processes have resulted in serious consequences to global natural resources (e.g. compaction, inorganic/organic contamination, and diminished microbial activity/diversity). The area under the affected soils continues to increase each year due to introduction of irrigation in new areas.

Salinization is recognized as the main threats to environmental resources and human health in many countries, affecting almost 1 billion ha worldwide/globally representing about 7% of earth's continental extent, approximately 10 times the size of a country like Venezuela or 20 times the size of France.

Impact of Salinity on Plants

Agricultural crops exhibit a spectrum of responses under salt stress. Salinity not only decreases the agricultural production of most crops, but also, effects soil physicochemical properties, and ecological balance of the area. The impacts of salinity include—low agricultural productivity, low economic returns and soil erosions. Salinity effects are the results of complex interactions among morphological, physiological, and biochemical processes including seed germination, plant growth, and water and nutrient uptake. Salinity affects almost all aspects of plant development including: germination, vegetative growth and reproductive development. Soil salinity imposes ion toxicity, osmotic stress, nutrient (N, Ca, K, P, Fe, Zn) deficiency and oxidative stress on plants, and thus limits water uptake from soil. Soil salinity significantly reduces plant phosphorus (P) uptake because phosphate ions precipitate with Ca ions. Some elements, such as sodium, chlorine, and boron, have specific toxic effects on plants. Excessive accumulation of sodium in cell walls can rapidly lead to osmotic stress and cell death. Plants sensitive to these elements may be affected at relatively low salt concentrations if the soil contains enough of the toxic element. Because many salts are also plant nutrients, high salt levels in the soil can upset the nutrient balance in the plant or interfere with the uptake of some nutrients. Salinity also affects photosynthesis mainly through a reduction in leaf area, chlorophyll content and stomatal conductance, and to a lesser extent through a decrease in photosystem II efficiency (Netondo et al., 2004). Salinity adversely affects reproductive development by inhabiting microsporogenesis and stamen filament elongation, enhancing programed cell death in some tissue types, ovule abortion and senescence of fertilized embryos. The saline growth medium causes many adverse effects on plant growth, due to a low osmotic potential of soil solution (osmotic stress), specific ion effects (salt stress), nutritional imbalances, or a combination of these factors. All these factors cause adverse effects on plant growth and development at physiological and biochemical levels, and at the molecular level.

In order to assess the tolerance of plants to salinity stress, growth or survival of the plant is measured because it integrates the up- or down-regulation of many physiological mechanisms occurring within the plant. Osmotic balance is essential for plants growing in saline medium. Failure of this balance results in loss of turgidity, cell dehydration and ultimately, death of cells. On the other hand, adverse effects of salinity on plant growth may also result from impairment of the supply of photosynthetic assimilates or hormones to the growing tissues. Ion toxicity is the result of replacement of K^+ by Na^+ in biochemical reactions, and Na^+ and Cl^- induced conformational changes in proteins. For several enzymes, K^+ acts as cofactor and cannot be substituted by $Na+$. High K^+ concentration is also required for binding tRNA to ribosomes and thus protein synthesis. Ion toxicity and osmotic stress cause metabolic imbalance, which in turn leads to oxidative stress. The adverse effects of salinity on plant development are more profound during the reproductive phase. Wheat plants stressed at 100–175 mM NaCl showed a significant reduction in spikelets per spike, delayed spike emergence and reduced fertility, which results in poor grain yields. However, Na^+ and Cl^- concentrations in the shoot apex of these wheat plants were below 50 and 30 mM, respectively, which is too low to limit metabolic reactions. Hence, the adverse effects of salinity may be attributed to the salt-stress effect on the cell cycle and differentiation. Salinity arrests the cell cycle transiently by reducing the expression and activity of cyclins and cyclin-dependent kinases that results in fewer cells in the meristem, thus limiting growth. The activity of cyclin-dependent kinase is diminished also by post-translational inhibition during salt stress. Recent reports also show that salinity adversely affects plant growth and development, hindering seed germination, seedling growth, enzyme activity, DNA, RNA, protein synthesis and mitosis.

Amelioration of Salinity

Salinization can be restricted by leaching of salt from root zone, changed farm management practices and use of salt tolerant plants. Irrigated agriculture can be sustained by better irrigation practices such as adoption of partial root zone drying methodology, and drip or micro-jet irrigation to optimize use of water. The spread of dry land salinity can be contained by reducing the amount of water passing beyond the roots. This can be done by re-introducing deep rooted perennial plants that continue to grow and use water during the seasons that do not support annual crop plants. This may restore the balance between rainfall and water use, thus preventing rising water tables and the movement of salt to the soil surface. Farming systems can change to incorporate perennials in rotation with annual crops (phase farming), in mixed plantings (alley farming, intercropping), or in site-specific plantings (precision farming). Although the use of these approaches to sustainable management can ameliorate yield reduction under salinity stress, implementation is often limited because of cost and availability of good water quality or water resource. Evolving efficient, low cost, easily adaptable methods for the abiotic stress management is a major challenge. Worldwide, extensive research is being carried out, to develop strategies to cope with abiotic stresses, through development of salt and drought tolerant varieties, shifting the crop calendars, resource management practices etc.

Use of Salt Tolerant Crops and Transgenics

Using the salt-tolerant crops is one of the most important strategies to solve the problem of salinity. Tolerance will be required for the "de-watering" species, but also for the annual crops to follow, as salt will be left in the soil when the water table is lowered. Salt tolerance in crops will also allow the more effective use of poor quality irrigation water. To increase the plant salt-tolerance, there is a need for understanding the mechanisms of salt limitation on plant growth and the mechanism of salt tolerance at the whole-plant, organelle, and molecular levels. Under saline conditions, there is a change in the pattern of gene expression, and both qualitative and quantitative changes in protein synthesis. Although it is generally agreed that salt stress brings about quantitative changes in protein synthesis, there is some controversy as to whether salinity activates specialized genes that are involved in salt stress. Salt tolerance does not appear to be conferred by unique gene(s). When a plant is subjected to abiotic stress, a number of genes are turned on, resulting in increased levels of several metabolites and proteins, some of which may be responsible for conferring a certain degree of protection to these stresses. Efforts to improve crop performance by transgenic approach under environmental stresses have not been that fruitful because the fundamental mechanisms of stress tolerance in plants remain to be completely understood.

Development of salt-tolerant crops has been a major objective of plant breeding programs for decades in order to maintain crop productivity in semiarid and saline lands. Although several salt-tolerant varieties have been released, the overall progress of traditional breeding has been slow and has not been successful as only few major determinant genetic traits of salt tolerance have been identified. 25 years ago Epstein et al. described the technical and biological constraints to solving the problem of salinity. Although there has been some success with technical solutions to the problem, the biological solutions have been more difficult to develop because a pre-requisite for the development of salt tolerant crops is the identification of key genetic determinants of stress tolerance. The existence of salt-tolerant plants (halophytes) and differences in salt tolerance between genotypes within salt-sensitive plant species (glycophytes) indicates that there is a genetic basis to salt response. Although a lot of approaches have been done for development of salt tolerant plants by transgenics complete success is not achieved yet. The assessment of salt tolerance in transgenic experiments has been mostly carried out using a limited number of seedlings or mature plants in laboratory experiments. In most of the cases, the experiments were carried out in greenhouse conditions where the plants were not exposed to those conditions that prevail in high-salinity soils (e.g. alkaline soil pH, high diurnal temperatures, low humidity, and presence of other sodic salts and elevated concentrations of selenium and/or boron). The salt tolerance of the plants in the field needs to be evaluated and, more importantly, salt tolerance needs to be evaluated as a function of yield. The evaluation of field performance under salt stress is difficult because of the variability of salt levels in field conditions and the potential for interactions with other environmental factors, including soil fertility, temperature, light intensity and water loss due to transpiration. Evaluating tolerance is also made more complex because of variation in sensitivity to salt during the life cycle. For example, in rice, grain yield is much more affected by salinity than in vegetative growth. In tomato, the ability of the plants to germinate under conditions of high salinity is not always correlated with the ability of the plant to grow under salt stress because both are controlled by different mechanisms, although some genotypes might display similar tolerance at germination and during vegetative growth. Therefore, the assessment of stress tolerance in the laboratory often has little correlation to tolerance in the field. Although there have been many successes in developing stress-tolerant transgenics in model plants such as tobacco, Arabidopsis or rice, there is an urgent need to test these successes in other crops. There are several technical and financial challenges associated with transforming many of the crop plants, particularly the monocots. First, transformation of any monocot other than rice is still not routine and to develop a series of independent homozygous lines is costly, both in terms of money and time. Second, the stress tolerance screens will need to include a field component because many of the stress tolerance assays used by basic researchers involve using nutrient-rich media (which in some cases include sucrose). This type of screen is unlikely to have a relationship to field performance. Third, because saline soils are often complex and can include NaCl, $CaCl_2$, $CaSO_4$, $Na2SO_4$, high boron concentrations and alkaline pH, plants that show particular promise will eventually have to be tested in all these environments.

Microbes: Abiotic Stress Alleviation Tool in Crops

Several strategies have been developed in order to decrease the toxic effects caused by high salinity on plant growth, including plant genetic engineering, and recently the use of plant growth-promoting bacteria (PGPB). The role of microorganisms in plant growth promotion, nutrient management and disease control is well known and well established. These beneficial microorganisms colonize the rhizosphere/endorhizosphere of plants and promote growth of the plants through various direct and indirect mechanisms. Previous studies suggest that utilization of PGPB has become a promising alternative to alleviate plant stress caused by salinity and the role of microbes in the management of biotic and abiotic stresses is gaining importance. The subject of PGPR elicited tolerance to abiotic stresses has been reviewed recently.

The term Induced Systemic Tolerance (IST) has been proposed for PGPR-induced physical and chemical changes that result in enhanced tolerance to abiotic stress. PGPR facilitate plant growth indirectly by reducing plant pathogens, or directly by facilitating the nutrient uptake through phytohormone production (e.g. auxin, cytokinin and gibberellins), by enzymatic lowering of plant ethylene levels and/or by production of siderophores. It has been demonstrated that inoculations with AM (arbuscular mycorrhizal) fungi improves plant growth under salt stress. Kohler et al., 2006 demonstrated the beneficial effect of PGPR Pseudomonas mendocina strains on stabilization of soil aggregate. The three PGPR isolates P. alcaligenes PsA15, Bacillus polymyxa BcP26 and Mycobacterium phlei MbP18 were able to tolerate high temperatures and salt concentrations and thus confer on them potential competitive advantage to survive in arid and saline soils such as calcisol. Kohler et al., 2009 investigated the influence of inoculation with a PGPR, P. mendocina, alone or in combination with an AM fungus, Glomus intraradices or G. mosseae on growth and nutrient uptake and other physiological activities of Lactuca sativa affected by salt stress. The plants inoculated with P. mendocina had significantly greater shoot biomass than the controls and it is suggested that inoculation with selected PGPR could be an effective tool for alleviating salinity stress in salt sensitive plants. Bacteria isolated from different stressed habitats possess stress tolerance capacity along with the plant growth-promoting traits and therefore are potential candidates for seed bacterization. When inoculated with these isolates, plants show enhanced root and shoot length, biomass, and biochemical levels such as chlorophyll, carotenoids, and protein. Investigations on interaction of PGPR with other microbes and their effect on the physiological response of crop plants under different soil salinity regimes are still in incipient stage. Inoculations with selected PGPR and other microbes could serve as the potential tool for alleviating salinity stress in salt sensitive crops. Therefore, an extensive investigation is needed in this area, and the use of PGPR and other symbiotic microorganisms, can be useful in developing strategies to facilitate sustainable agriculture in saline soils.

Alleviation of Abiotic Stress in Plants by Rhizospheric Bacteria

Besides developing mechanisms for stress tolerance, microorganisms can also impart some degree of tolerance to plants towards abiotic stresses like drought, chilling injury, salinity, metal toxicity and high temperature. In the last decade, bacteria belonging to different genera including Rhizobium, Bacillus, Pseudomonas, Pantoea, Paenibacillus, Burkholderia, Achromobacter, Azospirillum, Microbacterium, Methylobacterium, Variovorax, Enterobacter etc. have been reported to provide tolerance to host plants under different abiotic stress environments. Use of these microorganisms per se can alleviate stresses in agriculture thus opening a new and emerging application of microorganisms. Microbial elicited stress tolerance in plants may be due to a variety of mechanisms proposed from time to time based on studies done. Production of indole acetic acid, gibberellins and some unknown determinants by PGPR, results in increased root length, root surface area and number of root tips, leading to an enhanced uptake of nutrients thereby improving plant health under stress conditions. Plant growth promoting bacteria have been found to improve growth of tomato, pepper, canola, bean and lettuce under saline conditions.

Some PGPR strains produce cytokinin and antioxidants, which result in abscisic acid (ABA) accumulation and degradation of reactive oxygen species. High activities of antioxidant enzymes are linked with oxidative stress tolerance. Another PGPR strain, Achromobacter piechaudii ARV8 which produced 1-aminocyclopropane-1-carboxylate (ACC) deaminase, conferred IST against drought and salt in pepper and tomato. Many aspects of plant life are regulated by ethylene levels and the biosynthesis of ethylene is subjected to tight regulation, involving transcriptional and post-transcriptional factors regulated by environmental cues, including biotic and abiotic stresses. Under stress conditions, the plant hormone ethylene endogenously regulates plant homoeostasis and results in reduced root and shoot growth. In the presence of ACC deaminase producing bacteria, plant ACC is sequestered and degraded by bacterial cells to supply nitrogen and energy. Furthermore, by removing ACC, the bacteria reduce the deleterious effect of ethylene, ameliorating stress and promoting plant growth. The complex and dynamic interactions among microorganisms, roots, soil and water in the rhizosphere induce changes in physicochemical and structural properties of the soil. Microbial polysaccharides can bind soil particles to form microaggregates and macroaggregates. Plant roots and fungal hyphae fit in the pores between microaggregates and thus stabilize macroaggregates. Plants treated with Exo-poly saccharides (EPS) producing bacteria display increased resistance to water and salinity stress due to improved soil structure. EPS can also bind to cations including Na^+ thus making it unavailable to plants under saline conditions. Chen et al., 2007 correlated proline accumulation with drought and salt tolerance in plants. Introduction of proBA genes derived from B. subtilis into A. thaliana resulted in production of higher levels of free proline resulting in increased tolerance to osmotic stress in the transgenic plants. Increased production of proline along with decreased electrolyte leakage, maintenance of relative water content of leaves and selective uptake of K ions resulted in salt tolerance in Zea mays coinoculated with Rhizobium and Pseudomonas. Rhizobacteria inhabiting the sites exposed to frequent stress conditions, are likely to be more adaptive or tolerant and may serve as better plant growth promoters under stressful conditions. Moreover Yao et al., 2010 reported that inoculation with P. putida Rs 198 promoted cotton growth and germination under conditions of salt stress. Tank and Saraf showed that PGPRs which are able to solubilize phosphate, produce phytohormones and siderophores in salt condition promote growth of tomato plants under 2% NaCl stress.

In a study carried out by Naz et al., 2009, it was shown that strains isolated from Khewra salt range of Pakistan exhibited their tolerance when tested on saline media simulated by rhizosphere soil filtrate. Noteworthy, the isolates produced ABA in a concentration much higher than that of previous reports. Furthermore production of proline, shoot/root length, and dry weight was also higher in soybean plants inoculated with these isolates under induced salt stress. Likewise Upadhyay et al., 2011 studied the impact of PGPR inoculation on growth and antioxidant status of wheat under saline conditions and reported that co-inoculation with B. subtilis and Arthrobacter sp. could alleviate the adverse effects of soil salinity on wheat growth with an increase in dry biomass, total soluble sugars and proline content. Jha et al., 2011 reported that P. pseudoalcaligenes, an endophytic bacterium in combination with a rhizospheric B. pumilus in paddy was able to protect the plant from abiotic stress by induction of osmoprotectant and antioxidant proteins than by the rhizospheric or endophytic bacteria alone at early stages of growth. Plants inoculated with endophytic bacterium P. pseudoalcaligenes showed a significantly higher concentration of glycine betaine-like quaternary compounds and higher shoot biomass at lower salinity levels. While at higher salinity levels, a mixture of both P. pseudoalcaligenes and B. pumilus showed better response against the adverse effects of salinity. Nia et al., 2012 studied the effect of inoculation of Azospirillum strains isolated from saline or non-saline soil on yield and yield components of wheat in salinity and they observed that inoculation with the two isolates increased salinity tolerance of wheat plants; the saline-adapted isolate significantly increased shoot dry weight and grain yield under severe water salinity. The component of grain yield most affected by inoculation was grains per plant. Plants inoculated with saline-adapted Azospirillum strains had higher N concentrations at all water salinity levels.

Sadeghi et al., 2012 studied the plant growth promoting activity of an auxin and siderophore producing isolate of Streptomyces under saline soil conditions and reported increases in growth and development of wheat plant. They observed significant increases in germination rate, percentage and uniformity, shoot length and dry weight compared to the control. Applying the bacterial inocula increased the concentration of N, P, Fe and Mn in wheat shoots grown in normal and saline soil and thus concluded that Streptomyces isolate has potential to be utilized as biofertilizers in saline soils. More recently Ramadoss et al., 2013 studied the effect of five plant growth promoting halotolerant bacteria on wheat growth and found that inoculation of those halotolerant bacterial strains to ameliorate salt stress (80, 160 and 320 mM) in wheat seedlings produced an increase in root length of 71.7% in comparison with uninoculated positive controls. In particular, Hallobacillus sp. and B. halodenitrificans showed more than 90% increase in root elongation and 17.4% increase in dry weight when compared to uninoculated wheat seedlings at 320 mM NaCl stress indicating a significant reduction of the deleterious effects of NaCl. These results indicate that halotolerant bacteria isolated from saline environments have potential to enhance plant growth under saline stress through direct or indirect mechanisms and would be most appropriate as bioinoculants under such conditions. The isolation of indigenous microorganisms from the stress affected soils and screening on the basis of their stress tolerance and PGP traits may be useful in the rapid selection of efficient strains that could be used as bioinoculants for stressed crops.

References

- Consequences-soil-pollution-2558: sciencing.com, Retrieved 25 June, 2019

- Diseases-caused-by-soil-pollution-12287109: healthyliving.azcentral.com, Retrieved 14 March, 2019

- Effects-of-soil-pollution, environmental-issues: indiacelebrating.com, Retrieved 23 May, 2019

- Acidification, soil-health: qld.gov.au, Retrieved 18 January, 2019

- Dodd I.C., Perez-Alfocea F. Microbial amelioration of crop salinity stress. J. Exp. Bot. 2012;63(9):3415–3428WW

Permissions

We would like to thank the editorial team for lending their expertise to make the book truly unique. They have played a crucial role in the development of this book. Without their invaluable contributions this book wouldn't have been possible. They have made vital efforts to compile up to date information on the varied aspects of this subject to make this book a valuable addition to the collection of many professionals and students.

This book was conceptualized with the vision of imparting up-to-date and integrated information in this field. To ensure the same, a matchless editorial board was set up. Every individual on the board went through rigorous rounds of assessment to prove their worth. After which they invested a large part of their time researching and compiling the most relevant data for our readers.

The editorial board has been involved in producing this book since its inception. They have spent rigorous hours researching and exploring the diverse topics which have resulted in the successful publishing of this book. They have passed on their knowledge of decades through this book. To expedite this challenging task, the publisher supported the team at every step. A small team of assistant editors was also appointed to further simplify the editing procedure and attain best results for the readers.

Apart from the editorial board, the designing team has also invested a significant amount of their time in understanding the subject and creating the most relevant covers. They scrutinized every image to scout for the most suitable representation of the subject and create an appropriate cover for the book.

The publishing team has been an ardent support to the editorial, designing and production team. Their endless efforts to recruit the best for this project, has resulted in the accomplishment of this book. They are a veteran in the field of academics and their pool of knowledge is as vast as their experience in printing. Their expertise and guidance has proved useful at every step. Their uncompromising quality standards have made this book an exceptional effort. Their encouragement from time to time has been an inspiration for everyone.

The publisher and the editorial board hope that this book will prove to be a valuable piece of knowledge for students, practitioners and scholars across the globe.

Index

www.ingramcontent.com/pod-product-compliance
Lightning Source LLC
Chambersburg PA
CBHW062003190326

41458CB00009B/2957